CULTURE AND COSMOS
http://www.CultureAndCosmos.org

Culture and Cosmos is published twice a year, in northern spring/summer and autumn/winter, in association with the Sophia Centre for the Study of Cosmology in Culture, University of Wales Trinity Saint David.

Contributions and editorial correspondence should be addressed to:
Editors@cultureandcosmos.org

Editor: Dr. Nicholas Campion, the Editor of *Culture and Cosmos*, School of Archaeology, History and Anthropology, University of Wales Trinity Saint David, Lampeter, Ceredigion, Wales, SA48 7ED, UK.
E Mail **n.campion@uwtsd.ac.uk**

Subscriptions:
For two issues: Individuals £20*
Institutions £40
http://www.cultureandcosmos.org/subscription.html

Payment for hard copy is in line via paypal. For bank transfer apply to the editor.
*Members of the British Astronomical Association, The Astrological Association and The Historical Association are entitled to a discount. Please enquire.

Contributors Guidelines: Please see http://www.cultureandcosmos.org/submissions.html
.

Front cover: Geoff MacEwen, *Purgatorio* Canto 1 (*The Reed Bed*). See p. 59.

Published by Culture and Cosmos, School of Archaeology, History and Anthropology, University of Wales Trinity Saint David, Lampeter, Ceredigion, Wales, SA48 7ED, UK.

Printed by Lightning Source.

The Sophia Centre

http://www.uwtsd.ac.uk/sophia/

The Centre for the Study of Cosmology in Culture is an academic centre within the School of Archaeology, History and Anthropology and the Faculty of Humanities and the Performing Arts at the University of Wales Trinity Saint David.

The Centre's academic goals are

- 'to pursue research, scholarship and teaching in the relationship between astrological, astronomical and cosmological beliefs and theories, and society, politics, religion and the arts, past and present' and
- 'to undertake the academic and critical examination of astrology and its practice'.

The Centre's wider goal is stated in its title – to 'study cosmology in culture'. In a traditional sense, a cosmology is a worldview, an understanding of the cosmos which informs individual and social action and ideology. The Centre promotes research in the subject area, holds seminars and conferences, publishes scholarly material, is associated with Sophia Centre Press and supervises PhD students.

The Centre's teaching is focused on the MA Cultural Astronomy and Astrology. For further information see
http://www.uwtsd.ac.uk/ma-cultural-astronomy-astrology/

CULTURE AND COSMOS

www.CultureAndCosmos.org

Editor Nicholas Campion

Vol. 18 No. 1 Spring/Summer 2014 ISSN 1368-6534

Published in Association with
The Sophia Centre for the Study of Culture in Cosmology,
University of Wales Trinity Saint David
http://www.uwtsd.ac.uk/sophia/
School of Archaeology, History and Anthropology
http://www.uwtsd.ac.uk/aha/

Editorial

The Universe as Uncertainty and Action

A current widely-used astronomy textbook by Roger Freedman, Robert Geller and William Kaufmann declares that 'it is this [astronomers'] search for understanding that makes science more than merely a collection of data, and elevates it to one of the great adventures of the human mind'.[1] In partial support of this claim the authors quote Einstein's statement that 'The eternal mystery of the world is its comprehensibility'.[2] A search of the web reveals a number of sites which discuss what Einstein meant by this statement, although usually without coming to a particular conclusion. However, a check of the original source, Einstein's essay 'Physics and Reality', reveals that he did not actually directly write the words attributed to him as his opinion.[3] What he really wrote was slightly different and made careful use of quotation marks: 'One may say "the eternal mystery of the world is its comprehensibility"'.[4] By carefully enclosing the statement on comprehensibility in double quotation marks, he suggested that, while one may say that the world is comprehensible, equally one may say that it

[1] Roger A. Freedman, Robert M. Geller and William J. Kaufmann III, *Universe*, tenth edition (Basingstoke: W.H. Freeman and Company, 2014), p. 15.

[2] Freedman, Geller and Kaufmann, *Universe*, p. 15.

[3] Albert Einstein, 'Physics and Reality', 1936, in Albert Einstein, *Out of My Later Years* (New York: Philosophical Library, 1956), pp. 62-104.

[4] Einstein, 'Physics and Reality', p. 64.

is not. He also deliberately referenced the language of revealed religion by describing 'the fact that it [the universe] is comprehensible is a miracle'.[5]

The context for Einstein's article was the extent to which the previously rigid foundations of science had become problematic under the twin assault of relativity and quantum mechanics. Einstein's text, though, is similarly problematic, and he attempted simultaneously to assert the right of the physicist to comment on a range of disciplines in which they have no expertise, such as philosophy and psychology, but also implied, by contrast, that the entire universe is *not* comprehensible: even though he insisted that the material world can be investigated, he also conceded that 'we shall never understand' our ability to construct some kind of order in our understanding of the world'.[6] There is therefore a limit to what we can comprehend; the world is comprehensible but not understandable. This paradox runs through Einstein's argument. For example, he argues that a set of rules for investigation the universe must be stated, and they must be rigid in order for investigation to proceed. Yet, at the same time, these rules of investigation can only ever be arbitrary and provisional - 'the fixation will never be final', he wrote.[7]Einstein was emotionally attached to the idea that the universe is completely comprehensible, yet clearly struggling with the uncertain consequences of relativity and quantum mechanics.

Freedman, Geller and Kaufmann acknowledge the ambivalent nature of Einstein's statement on comprehensibility. The three authors conclude that 'It [the search for understanding of the universe] will continue as long as there are mysteries in the universe'.[8] The universe is therefore comprehensible in theory, but we do not comprehend it now.

The problem of the limits to the comprehensibility of the universe has a long literary lineage. In 1844 Robert Chambers included the following words in his seminal work on evolution:

> The mind fails to form an exact notion of a portion of space so immense; but some faint idea of it may be obtained from the fact, that, if the swiftest race-horse ever known had begin to traverse it, at full

[5] Einstein, 'Physics and Reality', p. 64.
[6] Einstein, 'Physics and Reality', p. 64.
[7] Einstein, 'Physics and Reality', p. 64.
[8] Freedman, Geller and Kaufmann, *Universe*, p. 15.

> speed, at the time of Moses, he would only as yet accomplished half his journey.[9]

Chambers' contemporary, the philosopher and historian Thomas Carlyle had a solution to the problem of the unknowability of the universe. Even though we can never know everything, at least we can act in the here and now, liberating ourselves and reforming society, making a better world. In his first great essay, 'Signs of the Times', Carlyle wrote,

> On the whole, as this wondrous planet, Earth, is journeying with its fellows through Infinite Space, so are the wondrous destinies embarked on it journeying through Infinite Time, under a higher guidance than ours. For the present, as our astronomy informs us, its path lies towards *Hercules*, the constellation of *Physical Power*: but that is not our pressing concern. Go where it will, the deep Heaven will be around it. Therein let us have hope and sure faith. To reform a world, to reform a nation, no wise man will undertake: and all but foolish men know, that the only solid, though a far slower reformation, is what each begins and perfects on himself.[10]

'Our grand business', Carlyle concluded, 'undoubtedly is, not to see what lies dimly at a distance, but to *do* what lies clearly at hand'.[11]

The Current Issue

This issue of *Culture and Cosmos* contains a range of articles spanning the journal's remit - Cultural Astronomy and Astrology. César Esteban explores theoretical issues in archaeoastronomy, drawing on his personal experience as an astrophysicist who has engaged with archaeology and arguing for the importance of landscape archaeology within archaeoastronomical research. Ronald Hutton's paper takes an entirely different perspective. As a historian, Hutton examines the extent to which

---

[9] Chambers, Robert, *Vestiges of the Natural History of Creation and Other Evolutionary Writings* (1844; Chicago: University of Chicago Press, 1994), pp. 1-2.

[10] Thomas Carlyle, 'Signs of the Times', *Edinburgh Review* 49 (1829), p. 441. See also Lawrence Poston, 'Millites and Millenarians: The Context of Carlyle's "Signs of the Times"', *Victorian Studies* 26, no. 4 (Summer 1983), pp. 381-406, p. 406; also see http://www.victorianweb.org/authors/carlyle/signs1.html [accessed 18 Juky2014].

[11] Carlyle, 'Signs of the Times', p. 406.

modern historians have projected their assumptions on to the past, finding, for example, evidence for the worship of a mother Goddess in British megalithic culture. Esteban is an astrophysicist and Hutton is a historian but they share an insistence on critical rigour and a rejection of comfortable assumptions.

Nick Kollerstrom represents the history of astrology, examining Galileo's connection with an apocalyptic astrological text. Clive Davenhall explores an equally curious footnote in the history of modern astronomy: the activities of the German showman Dr Katterfelto, who claimed to have made astronomical discoveries from his balloon. Lastly we include an interview with the distinguished print-maker and painter Geoff MacEwan, following an exhibition of his work inspired by Dante's *Divine Comedy* at the Christ Church Picture Gallery in Oxford.

Dr Nicholas Campion,
Sophia Centre for the Study of Cosmology in Culture,
School of Archaeology, History and Anthropology,
University of Wales Trinity Saint David.

Struggling for Interdisciplinarity: Reflections of an Astrophysicist Working in Cultural Astronomy[1]

César Esteban

Abstract: I present a personal view `on the role of astrophysicists and astronomers doing research in cultural astronomy. First, I discuss the definition of archaeoastronomy or cultural astronomy and its controversial interdisciplinary nature. I comment about the actual curricular problem of astrophysicists working in this topic and the difficult communication between astrophysicists—as well as other natural scientists—and archaeologists or anthropologists. I highlight the importance of accuracy in determining the orientation when mapping archaeological sites. Finally, I insist on the necessity of considering the celestial sphere as a part of the context of the archaeological sites, and that archaeoastronomy should be considered as a part of landscape archaeology.

A Definition Problem

The systematic study of astronomical orientations in Neolithic megalithic monuments began just over 50 years ago in the British Isles with the work of Gerald S. Hawkins on Stonehenge and the detailed studies of stone circles by Alexander S. Thom.[2] Hawkins was an astronomer and Thom an engineer and therefore used a rather different methodology to that familiar to archaeologists. These early researchers interpreted the existence of astronomical alignments as a demonstration of the high degree of geometric and astronomical knowledge gained by Megalithic man. In fact,

---

[1] First published as César Esteban, 'La astronomía cultural ¿es inter-disciplinar? Reflexiones de un astrofísico', *Complutum* Vol. 20, no. 2 (2009): pp. 69-77.

[2] Gerard S. Hawkins, 'Stonehenge decoded', *Nature* 200 (1963): pp. 306-8; Gerard S. Hawkins, 'Stonehenge: a Neolithic computer', *Nature* 202 (1964): pp. 1258-1261; Gerard S. Hawkins and John B. White, *Stonehenge decoded* (New York: Doubleday, 1963); Alexander Thom, *Megalithic sites in Britain* (Oxford: Clarendon Press, 1967); Alexander Thom, *Megalithic lunar observatories* (Oxford: Clarendon Press, 1971).

César Esteban,, 'Struggling for Interdisciplinarity: Reflections of an Astrophysicist Working in Cultural Astronomy', *Culture and Cosmos,* Vol. 18 no 1, Spring/Summer 2014, pp. 5-19.
www.CultureAndCosmos.org

for both, these sites functioned as true observatories.[3] Hawkins and White called the study of astronomical orientations astro-archaeology, defining it as an auxiliary discipline of archaeology, anthropology and history.[4]

Archaeologists harshly criticized early works on the orientation of megalithic monuments from both the conceptual and methodological points of view. In fact, a lively discussion was opened in *Antiquity*, one of the leading international archaeological journals.[5] According to S. Iwaniszewski, this disagreement between archaeologists and the first astro-archaeologists was mainly due to ‘the lack of a theory for analysing the astronomical knowledge in prehistoric societies’.[6] That is, the lack of a cultural, social and anthropological framework with which to interpret the data provided by astronomical alignments. Euan MacKie proposed a first but controversial attempt to explore the social implications of the results of astro-archaeological research on megalithic monuments.[7]

The precise name one uses for our topic of interest is not a trivial issue for many researchers. The term astro-archaeology was soon no longer used, at least in scientific circles. MacKie introduced the term ‘archaeoastronomy’ and defined it as the study of astronomical practices in the past.[8] On the other hand, in the early 1980s Anthony Aveni defined astro-archaeology as a hard branch of archaeoastronomy only limited to fieldwork and the subsequent calculations, detached from any kind of social or cultural analysis of the data.[9] Today pseudo-scientists have appropriated the term astro-archaeology so, for many of us, the situation of astro-archaeology is rather analogous to that of astrology with respect to astronomy. I recommend the book by John Michell about the early years of

---

[3] Thom, *Megalithic lunar observatories*, p. 9; Fred Hoyle, ‘Stonehenge—an eclipse predictor’, *Nature* 211 (1966): pp. 454-56.

[4] Hawkins and White, *Stonehenge decoded*, p. 121.

[5] Richard J. C. Atkinson, ‘Moonshine on Stonehenge’, *Antiquity* 40 (1966): pp. 212-16; Jacquetta Hawkes, ‘God in the Machine’, *Antiquity* 41 (1967): pp. 174-80; Fred Hoyle, ‘Speculations on Stonehenge’, *Antiquity* 40 (1966): pp. 262-76.

[6] Stanislaw Iwaniszewski, ‘De la astroarqueología a la astronomía cultural’, *Trabajos de Prehistoria* 51, no. 2 (1994): pp. 5-20.

[7] Euan MacKie, *Science and Society in Prehistoric Britain* (London: Paul Elek, 1977).

[8] Euan MacKie, ‘The last word on archaeoastronomy?’, *Archaeoastronomy (BCA)* 4, no. 1 (1981): p. 6.

[9] Anthony F. Aveni, *Observadores del cielo en el México antiguo* (Mexico City: Fondo de Cultura Económica, 1991): p. 14.

the studies of astronomical alignments and the history of astro-archaeology.[10]

One of the most lucid—but somewhat lengthy—definitions of archaeoastronomy has been proposed by Ed C. Krupp:

> Archaeoastronomy is the interdisciplinary study of prehistoric, ancient, and traditional astronomies worldwide within their cultural context. It includes both written and archaeological records. It embraces calendrics; practical observation; sky lore and celestial myth; symbolic representation of celestial objects, concepts and events; astronomical orientation of tombs, temples, shrines, and urban centers; symbolic displays involving celestial phenomena in the natural environment; traditional cosmology; and ceremonial application of astronomical tradition.[11]

By including traditional astronomy and cosmology—those of living cultures—Krupp is implicitly considering ethnoastronomy as a part of archaeoastronomy, although many interpret them as separate disciplines.

According to Iwaniszewski the diverse views on the definition of archaeoastronomy can be divided into three groups:[12]

> a) Archaeoastronomy is an interdisciplinary field of research and, along with ethnoastronomy, represents a holistic approach to the study of astronomy in the past and in the present.
> b) Archaeoastronomy is a branch of the history of science.
> c) Archaeoastronomy is a part of anthropology.

Iwaniszewski also questions the consideration of archaeoastronomy as an interdisciplinary field of study. He argues that its methods of interpretation use concepts and models of anthropology and history, but not of modern

---

[10] John Michell, *A Little History of Astro-Archaeology: Stages in the Transformation of a Heresy* (London: Thames & Hudson, 1989).

[11] Ed C. Krupp, 'Archaeoastronomy', in John Lankford, ed., *History of Astronomy. An Encyclopedia* (New York and London: Garland Publishing, Inc., 1997): pp. 21-30.

[12] Stanislaw Iwaniszewski, 'Archaeoastronomy and cultural astronomy: methodological issues', *Archaeologia e astronomia: esperienze e prospettive future*, Atti dei Convegni Lincei 121 (Rome: Accademia Nazionale dei Lincei, 1995): pp. 17-26.

astronomy.[13] Juan A. Belmonte seems to share this idea, as he says that archaeoastronomy 'is not a research topic of modern Astrophysics, and does not provide further information to advance in our knowledge of the physical universe'.[14] Iwaniszewski concludes that an archaeoastronomer is a researcher trained in anthropology—though not necessarily an anthropologist—and interested in the study of the role of the sky and celestial bodies in past and present cultures.[15]

For some scholars, the alleged interdisciplinary nature of archaeoastronomy is or may become a problem. For example, Krupp indicates that this can lead to the production of superficial studies.[16] M. Zeilik argues that it is precisely its interdisciplinary character that is responsible for the generally negative reception that archaeoastronomy has had among archaeologists.[17]

In 1990, Iwaniszewski proposed a new term: cultural astronomy, defining it as the study of the relationships between man and astronomical phenomena within a cultural context. Although considered as a separate discipline, it would be composed of four sub-disciplines: archaeoastronomy, ethnoastronomy, history of astronomy and social astronomy.[18] C. L. N. Ruggles and N. J. Saunders also endorsed this definition, indicating that cultural astronomy is closely related to the three anthropological disciplines—cultural anthropology, archaeology and

---

[13] Iwaniszewski, 'Archaeoastronomy and cultural astronomy: methodological issues'.

[14] Juan A. Belmonte, 'La investigación arqueoastronómica', in José Lull, ed., *Trabajos de Arqueoastronomía. Ejemplos de África, América, Europa y Oceanía* (Gandía: Agrupación Astronómica de La Safor, 2006): pp. 41-79.

[15] Iwaniszewski, 'Archaeoastronomy and cultural astronomy: methodological issues'.

[16] Ed C. Krupp, 'A glance into the Smoking Mirror', in Ray A. Williamson, ed., *Archaeoastronomy in the Americas* (Los Altos: Ballena Press/Center for Archaeoastronomy, 1981): pp. 55-59.

[17] Michael Zeilik, 'One Approach to Archaeoastronomy: An Astronomer's View', *Archaeoastronomy (JCA)* 6, nos. 1-4 (1983): pp. 4-7.

[18] Stanislaw Iwaniszewski, 'Astronomy as a Cultural System', *Interdisciplinarni zsledvaniya* 18 (1991): pp. 282-88.

ethnohistory—and one of its main objectives is the creation of a rigorous methodology with which to integrate data from such diverse sources.[19]

The development of archaeoastronomy and cultural astronomy in Spain has been summarized in several works.[20] The first publications were the product of specific collaborations between archaeologists and astronomers to study particular archaeological sites and the series of studies published by Michael Hoskin on dolmens on the Iberian Peninsula.[21] Some of Hoskin's works were done in collaboration with Spanish researchers—mostly archaeologists—and have been published since 1994.[22] Since 1993, Juan A. Belmonte and I, both astrophysicists at the Instituto de Astrofísica de Canarias (IAC, in Tenerife, Canary Islands) have undertaken extensive work in cultural astronomy—working together and separately—and published numerous papers and books.[23] One important addition to our research group at the IAC was A. César González-García, who is now one of the leading Spanish researchers in the area. In the last 10 to 15 years, we have had several additions to the Spanish archaeoastronomical community: the interdisciplinary group at the Universidad Complutense de Madrid, led

---

[19] Clive L. N. Ruggles and Nicholas J. Saunders, 'The Study of Cultural Astronomy', in Clive L. N. Ruggles and Nicholas J. Saunders, eds., *Astronomies and Cultures* (Niwot: University Press of Colorado, 1993): pp. 1-31.

[20] César Esteban, 'La arqueoastronomía en España', *Anuario del Observatorio Astronómico de Madrid para 2003* (Madrid: Instituto Geográfico Nacional, 2003): pp. 309-22; María L. Cerdeño, Gracia Rodríguez-Caderot, Pedro R. Moya, Ana Ibarra and Silvia Herrero, 'Los estudios de arqueoastronomía en España: estado de la cuestión', *Trabajos de Prehistoria* 63, no. 2 (2006): pp. 13-34.

[21] María S. López Plaza, Fernando Alonso, Manuel Cornide and A. Alvarez, 'Aplicación de la astronomía a la orientación de los sepulcros megalíticos de corredor en la zona noroccidental de la Península Ibérica', *Zephyrus* XLIV-XLV (1991-1992): pp. 183-192; Martín Almagro-Gorbea and Jean Gran-Aymerich, 'El estanque monumental de Bibracte', *Complutum Extra*, 1. (Madrid: Universidad Complutense de Madrid, 1991).

[22] See a compilation in Michael Hoskin, *Tombs, Temples and their Orientations: A New Perspective on Mediterranean Prehistory* (Bognor Regis: Ocarina Books, 2001).

[23] Juan A. Belmonte, *Las leyes del cielo* (Madrid: Temas de Hoy, 1999); Juan A. Belmonte, *Pirámides, templos y estrellas* (Barcelona: Crítica, 2011); Juan A. Belmonte and Michael Hoskin, *Reflejo del cosmos. Atlas de arqueoastronomía en el Mediterráneo antiguo* (Madrid: Equipo Sirius, 2002); Antonio Aparicio and César Esteban, *Las pirámides de Güímar. Mito y realidad* (Santa Cruz de Tenerife: Centro de la Cultura Popular Canaria, 2005).

by M. L. Cerdeño and G. Rodriguez, who have carried out interesting studies on Celtiberian culture; the work of M. Pérez Gutiérrez on Spanish Iron Age sites; and the studies carried out by Francisco Burillo Mozota and Maria P. Burillo on Celtiberian sites and ancient representations of constellations.[24]

A Curricular Problem

As discussed in the preceding section, for several scholars it would seem that astronomers have little role to play in cultural astronomy, at least in carrying out high-level studies. Honestly, I do not think this is the case. The historical development of our discipline has led to an evolution of its actors, as we have seen: first they were astronomers and engineers but gradually archaeologists and anthropologists have been taking a more important role. This is a reasonable and healthy evolution, because these professionals are precisely the ones most interested in the output of cultural astronomy studies. Moreover, archaeologists and anthropologists possess the training and methodologies with which archaeoastronomical results can be understood in their context. It is clear that astronomical alignments and other kinds of raw archaeoastronomical data are themselves not the important thing. They acquire full meaning when they help us to better understand the symbolism, religion and social and economical relations of human communities. Therefore, if an astronomer or astrophysicist wants to

---

[24] María L. Cerdeño, Gracia Rodríguez and Marta Folgueira, 'El paisaje funerario de la cultura celtibérica', *Anales de Prehistoria y Arqueología* 17-18 (2001-2002): pp. 177-85; María L. Cerdeño, María C. Hernández, Gracia Rodríguez and Marta Folgueira, 'Novedades culturales y metodológicas en la necrópolis de Herrería (Guadalajara)', *Novedades arqueológicas celtibéricas* (Madrid: Publicaciones del Museo Arqueológico Nacional, 2004), pp. 43-62; Gracia Rodríguez, María L. Cerdeño, Marta Folgueira and Teresa Sagardoy, 'Observaciones topoastronómicas en la Zona Arqueológica de El Ceremeño (Herrería, Guadalajara)', *Complutum* 17 (2006): pp. 133-43; Manuel Pérez Gutiérrez, 'Astronomía y Geometría en la Vettonia', *Complutum* 20, no. 2 (2009): pp. 141-64; Manuel Pérez Gutiérrez, Jordi Diloli Fons, David Bea Castaño and Samuel Sardà Seuma, 'Astronomy, culture and landscape in the Early Iron Age in the Ebro Basin', in Clive L. N. Ruggles, ed., *"Oxford IX" International Symposium on Archaeoastronomy*, Proceedings of the IAU Symposium No. 278 (Cambridge: Cambridge University Press, 2011): pp. 382-89; Francisco Burillo Mozota, María P. Burillo and Puy Segurado, 'De la investigación a la escuela: "Segedadenoche". Una reinterpretación teatralizada de la cosmogonía celtibérica', *Complutum* 20, no. 2 (2009): pp. 195-210.

do a valuable job in archaeoastronomy and cultural astronomy, she or he should acquire some serious training in the methodologies of the humanities.

From my professional experience and taking into account the current situation, the ideal researcher in cultural astronomy—to have higher possibilities to follow a consistent career—should be an archaeologist or anthropologist who acquires serious training in positional astronomy. A paradigmatic example of this would be the Mexican school of archaeoastronomy and cultural astronomy, such as the research groups organized around J. Broda (Instituto de Investigaciones Históricas, in the Universidad de Nacional Autonóma de México [UNAM]) and S. Iwaniszewski (Escuela Nacional de Antropología e Historia), both in Mexico City. These groups develop research projects and train new professionals in cultural astronomy from the fields of archaeology and anthropology. In Spain, we do not have a long tradition in interdisciplinary groups working in cultural astronomy, although those led by M. L. Cerdeño at the Universidad Complutense de Madrid and F. Criado-Boado and A. C. González-García at the Instituto de Ciencias del Patrimonio in Santiago de Compostela are seminal in this regard.

During my years of experience in teaching positional astronomy in the Degree on Physics at the Universidad de La Laguna, I have noticed that cultural astronomy arouses much interest among students. Every year, I had undergraduate students interested in carrying out works of introduction to research in this field. In fact, the level of interest on cultural astronomy is fairly similar to other topics I have taught at the University, the physics of interstellar matter or exobiology. When an undergraduate of astrophysics would ask me to work in cultural astronomy, I use to be frank and say that the experience gained with this kind of work can hardly have a place in the current design of the *curriculum vitae* of an astrophysicist. At any rate, they usually complete the work successfully and are happy to do it. In Spain, there is an extremely limited number of astrophysicists who have done or are doing a PhD in cultural astronomy and there are no contract offers for postdoctorals in this area, so that the academic and professional training in this topic is extremely difficult, at least until now, for astrophysicists. However, the National Plan for Astronomy and Astrophysics of the Spanish Government made a breakthrough in this direction by approving—in three consecutive occasions—three-year research projects devoted to cultural astronomy led by J. A. Belmonte. This was an official recognition of this branch of research by the rest of Spanish astronomers. Unfortunately, these research projects were never endowed with postdoctoral contracts but only with PhD studentships.

In my particular case, I dedicate most of my research to astrophysical issues. The work in cultural astronomy occupies a limited percentage of my time—a varying percentage depending on the circumstances and commitments of each moment. I started working on astronomical alignments shortly before finishing my PhD thesis, which was devoted to the determination and analysis of the chemical composition of nebulae associated with evolved massive stars. Some years later I got a permanent position as lecturer in the Department of Astrophysics at the Universidad de La Laguna, with a research and teaching profile in physics of the interstellar medium. To gain that position, I defended the merits of my curriculum vitae in a public exam before a panel of five astrophysicists. Of course, I mentioned the work I had done in cultural astronomy, but only as an additional merit. Honestly, I think this was the most reasonable strategy to get the position, taking into account its profile and the composition of the panel. Fortunately, getting a permanent position has allowed me to work in cultural astronomy without pressure, but the problems of astrophysicists who like to do other kinds of research does not stop here. In Spain, those who have permanent positions at universities or research centres have the possibility to obtain a complement to their salaries based on the evaluation of the quantity and quality of their research activities during six-years periods (the so-called *sexenios*). As I have to apply to the physics and astronomy panel, the merits presented for evaluation should be related to astrophysical research. The presentation of merits based on research in cultural astronomy may very probably be undervalued or even ignored by the panel, so one needs to do research on astrophysics to obtain such *sexenios*. Unfortunately, the situation of a researcher with a postdoctoral position or a temporary research contract—currently, the situation of most Spanish researchers under the age of 40—is far more difficult. The enormous competitiveness and the lack of positions and research contracts in cultural astronomy make it extremely difficult to build a career in this topic. Disseminating the results of our research among our colleagues is a key issue in getting any kind of official recognition by the astrophysical community. This is a true challenge for our still small group. The ideal situation would be to reach a certain critical mass of staff researchers with well-funded projects in cultural astronomy to ensure some representation and appreciation by the community.

Seeking Recognition by Archaeologists

The controversies that appeared with the early astro-archaeological works resulted in a rather general rejection of the discipline by the archaeological community. This has been—and still is—a handicap on the recognition of cultural astronomy as a valid field of research for many archaeologists. However, the main problem is the continuous development of a genuine pseudo-science under the heading of astro-archaeology, which confuses both the general public and the uninformed professionals.[25] An additional fact that does not contribute to the communication between astronomers and archaeologists is the important differences between the epistemology of the 'hard' or natural sciences and humanities, which may lead to mutual misunderstanding, prejudice and, in extreme cases, to a complete disregard. Cerdeño et al. discussed the lack of a common language between the communities, but fortunately noted some improvements: 'The scientism of the earlier authors (referring to the Spanish case), always astrophysicists, was completely exclusive, but that problem is becoming solved with the use of a more accessible language that allows easier access to technical data'.[26] The maturity that comes with experience, the dialogue—in some cases struggle—with journal referees, and above all, the contact and collaboration with archaeologists, have helped to improve the communication.

Sometimes, I feel that archaeoastronomical findings seem to have a rather limited impact on archaeological research, at least in the cultural areas more familiar to me. This makes me think that perhaps we are trying to answer questions that nobody has formulated or nobody is interested in.[27] While this may be true in some cases, I am convinced that this is a problem of the pioneering character of our research, because we are actually opening a new way of obtaining information about important cultural aspects of past societies. The main goal of cultural astronomy is to provide information about the ideological and religious systems of past cultures. For example, by studying the calendar we can infer aspects of the

---

[25] César Esteban, 'Arqueoastronomía y pseudociencia', in Inés Rodríguez Hidalgo, Luis F. Díaz Vilela, Carlos J. Álvarez González and José M. Riol Cimas, eds., *Ciencia y pseudociencias: realidades y mitos* (Madrid: Equipo Sirius, 2004), pp. 249-59; Belmonte, 'La investigación arqueoastronómica': pp. 41-79.

[26] Cerdeño et al., 'Los estudios de arqueoastronomía en España: estado de la cuestión'.

[27] I think this thought arose in a conversation with my colleague and friend Ivan Šprajc.

organization of the subsistence cycle and its relationship with the environment, which is often also intimately related to the social cycles. These cycles reflect the organization of the community at different levels and the power relationships. Fortunately, in Spain the number of archaeologists that are beginning to take into account these aspects is continuously increasing, and this is good news.[28] One example is the opinion of the renowned reputed archaeologist T. Chapa Brunet concerning archaeoastronomical works carried out on Iberian culture:

> Considering the data we have at hand, these kinds of proposals are for the moment rather speculative, but the knowledge of the annual rhythms of a society tells us a lot about its economic and ideological organization, and therefore more emphasis should be given to this type of documentation when studying the archaeological sites.[29]

The Celestial Sphere as a Context

One of the trends of modern archaeology is the study of sites in relation to their geographic and ecological contexts, which is now known as landscape archaeology. I. Grau Mira gives a lucid definition of this discipline: 'Landscape archaeology may be defined as a comprehensive and multidirectional analysis of the elements of the landscape that tries to

---

[28] López Plaza et al., 'Aplicación de la astronomía a la orientación de los sepulcros megalíticos de corredor en la zona noroccidental de la Península Ibérica'; Almagro-Gorbea and Gran-Aymerich, 'El estanque monumental de Bibracte'; Cerdeño et al., 'El paisaje funerario de la cultura celtibérica'; Cerdeño et al., 'Novedades culturales y metodológicas en la necrópolis de Herrería (Guadalajara)'; José L. Escacena Carrasco, 'Allas el estrellero, o Darwin en las sacristías', in José L. Escacena Carrasco and Eduardo Ferrer Albelda, eds., *Entre Dios y los hombres: el sacerdocio en la antigüedad*, Spal Monografías VII (Seville: Universidad de Sevilla, 2006), pp. 103-156; Felipe Criado and Marco García Quintela, 'Landscape, archaeology and ethno-astronomy: a union foretold', in Mauro P. Zedda and Juan A. Belmonte, eds., *Lights and Shadows in Cultural Astronomy* (Isili: Associazione Archeofila Sarda, 2007), pp. 87-99.

[29] Teresa Chapa Brunet, 'Sacrificio y sacerdocio entre los iberos', in José L. Escacena Carrasco and Eduardo Ferrer Albelda, eds., *Entre Dios y los hombres: el sacerdocio en la antigüedad*, Spal Monografías VII (Seville: Universidad de Sevilla, 2006), pp. 157-77.

understand the society that models the space and interacts with it'.[30] It seems obvious that the celestial sphere should be considered as part of the context for understanding the possible reasons for the location of any archaeological site. In fact, some archaeologists have begun to consider the study of the orientations of archaeological remains as a part of landscape archaeology.[31]

An example of an archaeological monument tied in context at different levels may be the Iberian temple of Tossal de Sant Miquel de Llíria, in the province of Valencia, Spain.[32] The building is embedded within the layout of an Iron Age Iberian settlement, but precisely in a place where its orientation is twisted to lie in an east-west direction and in a high place with an open view of the Eastern horizon. Curiously, the space in front of the temple entrance is clear of the remains of any other construction. The archaeologists that worked on the site confirmed that no other buildings were ever erected in that particular spot—at least during the existence of the Iberian settlement—leaving an empty space that allowed a view of the eastern horizon from the temple. This first fact indicates the importance of the location and arrangement of the temple within its nearest context, the settlement. As I have said, the building is oriented in the east-west direction and is facing very precisely the point of the horizon where the sun rises at the equinoxes. This second fact relates the temple with its celestial context. But the relations do not end here: seen from the temple, the sunrise at the equinoxes takes place on the top of a small mountain that breaks the monotony of flat southeast horizon. Therefore, the mountain could have a function as a marker or calendrical reference, providing a third fact that indicates the integration of the archaeological site within the geographical context.

Because of my training as astrophysics, the type of research I usually do in cultural astronomy is the determination and analysis of alignments defined in archaeological sites and the horizon that surrounds them. In most cases, I collaborate with archaeologists that have excavated the sites.

---

[30] Ignasi Grau Mira, *La organización del territorio en el área central de la Contestania Ibérica* (San Vicente del Raspeig: Publicaciones de la Universidad de Alicante, 2002), p. 20.

[31] Criado and García Quintela, 'Landscape, archaeology and ethno-astronomy: a union foretold'.

[32] César Esteban and Soraya Moret, 'Ciclos de tiempo en la cultura ibérica: la orientación astronómica en el templo del Tossal de Sant Miquel de Llíria', *Trabajos de Prehistoria* 53, no. 1 (2006): pp. 167-78.

This collaboration has proved extremely useful especially at two key moments of the research. Firstly during the fieldwork, where we try to combine our different but complementary points of view—the astronomical and archaeological— for better understanding the site and to consider subtle important aspects that otherwise could go unnoticed to one of us working alone. And secondly, in the discussion of the results where we try to integrate the archaeoastronomical findings into the wider ideological and archaeological context of the culture we are investigating.

The astronomical analysis of the alignments defined in a monument tells us if their spatial arrangement is related to the points of the horizon where the rising or setting of certain celestial bodies occur on the local horizon. The measurable elements we analyse are:

a) The orientation of different structures of the archaeological site.
b) The relation between elements of the horizon that surrounds the site and position of celestial bodies.

To carry out a valuable archaeoastronomical study, it is clear that one should measure the alignments defined in archaeological sites with the highest precision possible. In principle, one could do this kind of research simply working on published maps of them, but unfortunately this has been demonstrated to be a very risky task. My experience has shown that most of the published plans are not made with sufficient precision and therefore are misleading or even useless for our purposes. The general impression is that a significant amount of archaeologists are not particularly worried in accurately placing the sites in space, something that is essential for us. First, many published maps do not show the position of the north, and when it is indicated one does not know if it corresponds to magnetic or true north. This distinction is important because, in general, they do not coincide and the difference is usually several degrees. Magnetic north is the one that provides the compass point and its position relative to the true north changes with time and with the site coordinates. Using magnetic north in a map could be useful if the date of the planimetric measures is indicated in it, since the variation of the angular difference between magnetic and true north—what is called magnetic declination—over time can be estimated with some precision unless there are local magnetic anomalies.[33] However, the plans containing magnetic north never provide the date they were made. Using published maps of archaeological sites has provided me some surprises. There is a particular case of an Iberian temple in which the difference between the orientation of the north indicated in the plan and the true north I determined at the site was 42°, or nearly a half quadrant. An error of this magnitude clearly indicates that the layout of the building in space was of no particular concern to the archaeologists who published the map.

In recent years, the arrival of Google Earth has provided us with a tool that may be a powerful source of archaeoastronomical data. With this application we can determine alignments of architectural structures that are identified in the high-resolution images available. An advantage of Google Earth is that the images are orientated with respect to true north and, in principle, we can measure azimuths. Interesting examples of the use of this tool can be found in the study carried out by Belmonte on the orientation of temples of the Kingdom of Kush in Sudan, and also in the book by F. Herráiz Sánchez on the geometry and possible astronomical implications of the original layout of the city of San Cristóbal de La Laguna in Tenerife.[34] My experience with Google Earth is quite positive; for several sites, I have compared azimuth measurements of distant elements of the local horizon obtained with a precision compass or theodolite with those determined from the web application and the agreement is about one degree. Although Google Earth is a tool that can be very useful in many cases, it should never replace fieldwork at the site when it is feasible.[35] Obviously, satellite maps can never give the wealth of detail provided by fieldwork at the site. One extremely important drawback of using Google Earth is that it does not provide information on the precise shape and

[33] There are some web pages where the magnetic declination can be obtained, see for example at http://www.ngdc.noaa.gov/geomag/ [accessed 31 October 2014] or http://www.qibla.com.br [accessed 31 October 2014]. These variations may be terrific in special geological areas. For example, in volcanic terrain as in the Canary Islands the determination of magnetic declination at the site is mandatory due to the large local variations.

[34] Juan A. Belmonte, 'Kingdom of Kush', in Clive L. N. Ruggles, ed., *Handbook of Archaeoastronomy and Ethnoastronomy* (Heidelberg: Springer, 2014), pp. 1541-48; Fernando Herráiz Sanchez, *La Laguna oculta. El cielo y la piedra.* (Santa Cruz de Tenerife: Ediciones Idea, 2007).

[35] In some cases, Google Earth may be the only possibility to do archaeoastronomical work, as in the case of geographical areas where access is dangerous because of political situations or wars.

height of the different features of the local horizon—the visibility—which can be essential to correctly assign an astronomical relation to an alignment measured in an archaeological structure or to find an astronomical marker over the horizon. Another limitation of using Google Earth is that one can only measure relatively large structures that are distinguishable from the air at the spatial resolution of the images available. However, as almost everything can now be found in the web, one can get an idea of the local horizon from the armchair using the web application HeyWhatsThat.[36] It provides a scaled view of the horizon from a specified location, defined as a map point on Google Maps.[37] My own experience with HeyWhatsThat is rather limited, but the horizon views are of rather low resolution and its accuracy drops dramatically when the local horizon is close to the site and you do not have a precise determination of the height of the site above sea level.

Epilogue: Recreating the Ancient Skies

I want to end this article by discussing an experience that archaeoastronomy that may illustrate the discipline's use in archaeology. Sometimes, when working on astronomical orientations one finds that, from an archaeological site, a celestial object — generally the Sun or Moon — either transits or has its rising or setting at a conspicuous place and/or produces a striking phenomenon in a singular moment of its apparent orbital cycle around the Earth (solstices and equinoxes in the case of the Sun, lunastices in the case of the Moon). If the site has not too early a chronology, we may be lucky enough to observe the astronomical phenomenon practically in the same manner as did those who built the archaeological site. Among the places I have studied, the equinox sunrise at the Iberian shrine of El Amarejo or the equinox sunset at the cave–sanctuary of Castellar may be good examples of such a 'rendezvous with the past'.[38] The phenomena that occur on those two places are full of

---

[36] http://www.heywhatsthat.com [accessed 31 October 2014].

[37] For an example of the use of this application, see Andrea Rodríguez-Antón, Juan A. Belmonte and A. César González-García, 'Orientation of Roman Camps and Forts in Britannia', in Fabio Silva, Kim Malville, Tore Lomsdalen and Frank Ventura, eds., *The Materiality of the Sky, SEAC 2014* (Ceredigion, UK: Sophia Centre Press, 2015): in press.

[38] César Esteban, 'Elementos astronómicos en el mundo religioso y funerario ibérico', *Trabajos de Prehistoria* 59, no. 2 (2002): pp. 81-100; César Esteban,

symbolism, should have a public dimension, and may be interpreted as true hierophanies. It is difficult to explain what one feels when discovering an astronomical phenomenon of this type: it is like receiving a message from the distant past, reliving an experience after centuries or millennia of being forgotten. I suppose that an archaeologist must feel something similar when opening a grave, but the difference is that the archaeologist discovers something that is definitely dead while the eyes of the archaeoastronomer may see a phenomenon that, although intangible, is still alive, returning with the perfect accuracy of the celestial cycles. It is the generosity of heavens; one can understand why it was so important in ancient culture.

Carmen Rísquez and Carmen Rueda, 'An evanescent vision of the sacred? The equinoctial sun at the Iberian sanctuary of Castellar', *Mediterranean Archaeology and Archaeometry* (2014): in press.

Prehistoric British Astronomy: Whatever Happened to the Earth and Sun?

Ronald Hutton

Abstract: During the first half of the twentieth century, it was an orthodoxy among British archaeologists that the New Stone Age peoples of the island had worshipped an Earth Goddess, in chambered tombs, and then been conquered by foreigners who ushered in a Bronze Age, characterised by circular temples dedicated to a new religion focused on the heavens. In the second half of the century, belief in this sequence collapsed, and experts more or less abandoned attempts to reconstruct religion during this period of prehistory. At the same time it remains true that many of its monuments have clear alignments on heavenly bodies. What now, then, can be done to bring together this evidence with prevailing scholarly attitudes?

Anybody with only a passing acquaintance with the prehistoric monuments of the British Isles will know that many of them are aligned, often with great accuracy, upon heavenly bodies. Indeed, the very greatest and most famous of all are, in each one of the three historic realms of the British Isles. The most celebrated ancient site in Ireland is Newgrange, the huge Neolithic passage tomb in the Boyne Valley, the entrance of which is designed to face the midwinter sunrise. A small aperture above it admits a ray of light from the rising sun which travels down a passage behind to strike a carved stone at the back of the main chamber. It is a magnificent feat of engineering from around 3200 BCE.[1] The same effect is found at the single most famous and sophisticated Neolithic structure in Scotland, the Neolithic passage tomb at Maes Howe in Orkney. Built a couple of centuries later than Newgrange, it admits the midwinter sunset down its passage to light up the main chamber.[2]

By contrast, it is generally known that the entrance to the greatest prehistoric monument in England, which is the most famous in the entire

[1] Michael J. O'Kelly, *Newgrange* (London: Thames and Hudson, 1982).

[2] Patrick Ashmore, *Maes Howe* (Edinburgh: Her Majesty's Stationary Office, 2002).

Ronald Hutton, 'Prehistoric British Astronomy: Whatever Happened to the Earth and Sun?', *Culture and Cosmos*, Vol. 18 no 1 (Spring/Summer 2014), pp. 21-34.
www.CultureAndCosmos.org

world, is aligned on the midsummer sunrise. This is of course Stonehenge, which was built in its present form between 2600 and 2400 BCE: still a very long time ago, and contemporary with the earliest of the Egyptian pyramids. What is less well known, although it is becoming more so, is that Stonehenge once embodied a still more dramatic alignment upon the midwinter sunset, just as Maes Howe had done. The largest of all the freestanding three-stone settings in the centre, the Great Trilithon, was positioned so that the sun at midwinter set directly behind it. This caused the red light of the dying sun to pour through the narrow gap between the two uprights. It is not, I think, too fanciful to compare what would have been the effect with that of birth or menstrual blood flowing from between the legs of a giant female figure. I also concede, of course, that this is only one possible reading of it. The effect itself is long lost, because the builders made a tragic error in its construction. They found a single enormous stone to provide one of the uprights, which could be planted deep enough in the earth to secure it completely. It is actually the tallest prehistoric standing stone in the British Isles. They could not, however, find one to match it which was of equal length. So they cheated, by finding a stone of the right shape which was much shorter but had a piece projecting out horizontally from its end, like a shoe. They hoped that if the stone were put upright, with this projection sticking out under the topsoil, and jammed a big stone lintel on top to join it to the genuinely long stone next to it, all would be well. The shoe-like effect would provide some sort of anchor, while the lintel would push the stone down and fasten it to the stable one beside it, making the whole structure safe. This proved mistaken. At some unknown date, the shorter stone toppled over and broke, shedding its lintel, burying the altar stone, and rendering the centre of the monument unusable. It also, of course, ruined the whole effect of the midwinter sunset. At any rate, that effect was planned into the heart of Stonehenge's design and purpose, and was one of scores of alignments on the movements of the sun found in Britain's megalithic monuments.[3]

In the twentieth century, astronomers claimed to detect many more alignments focussed upon the moon and stars, though these have proved more controversial. It should be emphasised also that the undoubted solar connections are of various different kinds, and that most megaliths, even in

[3] It is discussed in most of the many recent books on Stonehenge, but my personal favourite among the references is from Julian Richards, *The Amazing Pop-Up Stonehenge* (Swindon: English Heritage, 2005), pp. 14-15, where it can be seen recreated in cardboard!

the same district, were not given any. This is not surprising, and should not be worrying, because it fits in with the usual pattern in Neolithic monuments of intense local diversity and creativity. Repeatedly a similar basic language of ceremonial architecture, found across the whole of the British Isles and often across Western Europe, was interpreted at local level in a very wide range of different forms.[4] I would emphasise here that I have no expertise in astronomy and my concern here is not with the question of how its relationship with prehistoric monuments *should be* interpreted. Instead, as a cultural historian, I am interested in *how* it has been interpreted; and that is a concern which fits well into those of this collection of essays. In a predecessor to it, I have considered the problem of why British archaeologists have not been more interested in the astronomical aspects of ancient sites, especially in recent years.[5] Here I am going to address a different aspect of the subject altogether: the manner in which they have written about the cosmological aspects of those sites. In other words, I am interested in the implications that they have drawn from their form, including their orientation on the sky. Those conclusions have altered significantly during my own lifetime, and the alterations concerned can in my opinion provide some very interesting insights into changes in modern British culture.

What should be emphasised here is that the three great monuments which I have discussed above—Newgrange, Maes Howe and Stonehenge—were all concentrated in the heyday of solar alignments for British and Irish monuments: between 3200 and 2400 BCE. Before and after that period, those interested in archaeoastronomy have found much less evidence of interest in the sun. Furthermore, the appearance of apparent intense interest in it is part of a much bigger change in British and Irish prehistory. It was no less than a revolution in attitudes to sacred space in the minds of prehistoric people. Put simply, the classic monument of the fourth millennium BCE, 4000 to 3000, is what I call the tomb-shrine. Local British names for it include long barrow, long cairn, dolmen, passage grave and cromlech. It is found all the way around Western Europe from Spain to Sweden. It consists of a stone or wooden chamber, usually containing human remains and often contained within a mound of earth or stones, much larger than was necessary merely to cover the chamber. These

[4] For this see Ronald Hutton, *Pagan Britain* (New Haven: Yale University Press, 2013), Chapters 2 and 3.

[5] Ronald Hutton, 'The Strange History of British Archaeoastronomy', *Journal of the Study of Religion, Nature and Culture* 7, no. 4 (2013): pp. 376-396.

structures were intended to make impressive statements in the landscape: they are indeed the first widely distributed form of monument in the story of humanity. Most of them have no clear alignments on the movements of heavenly bodies. Even within the same district, they face in more directions than are covered by the movements of the sun. Most would fall within those of the moon, but not all, and stars move around too much, and the dating of the monuments is too imprecise, for stellar alignments to be proved. The mounds and chambers are also of many different shapes, but forms of rectangle are the most common for both.[6]

In the years around 3000, however, the peoples of the British Isles became fascinated by round shapes for the first time, and the standard sacred unit of space became the circle for at least one and a half thousand years. At the same time, monuments began sometimes to be orientated, with great precision, on the movements of the sun, especially at the solstices. These changes took two different forms, however, in different halves of the archipelago. In Ireland, West Wales and Northern Scotland, it was grafted onto the older tradition of the tomb-shrine. The result was to take this tradition to its greatest achievements, in structures like Newgrange and Maes Howe: huge round mounds containing long passages leading to chambers with those alignments on the midwinter sun. In most of Britain, however, the tomb-shrines were abandoned, with no more being built and those still in use being blocked up. Instead, people took to holding ceremonies in open-air, circular enclosures, the materials of which depended on what the local geology provided. In areas of soft soil, they were made of earth, consisting of banks piled up around ditches: archaeologists call these structures henge monuments. In regions with plentiful timber, rings of wooden posts were erected, and where large stones were abundant, they were put up on end to form circles of megaliths. Stone circles are the classic surviving monuments of the third millennium BCE in Britain, after the earthen henges have been ploughed down and the timber rings rotted away. Many of the greatest of the new ceremonial landscapes, like those around Avebury and Stonehenge in Wiltshire, combined all of these forms: stone circles inside banks and ditches, often succeeding wooden structures or having those nearby. As said, some of them incorporated the new alignments on the sun. The dead were no longer put into the tomb-shrines, where their remains were a major

---

[6] See Ronald Hutton, *The Pagan Religions of the Ancient British Isles* (Oxford: Blackwell, 1991), pp. 16-51, revised and enlarged in Hutton, *Pagan Britain*, Chapter 2.

part of religious rites. Instead, they were sealed under smaller circular mounds, called round barrows in England. Increasingly, they were cremated before burial. Once buried, their bodies were cut off from the world of the living, but often accompanied by valuable goods, as they had not been in the tomb-shrines. These may have been for use in the next world, or gifts made to honour the dead, or possessions of the deceased, too strongly associated with them for others to feel safe using.[7] All this is undoubted archaeological fact. It represents a revolutionary change in the nature of the religious monuments of the British Isles, hinging on the transition between the fourth and third millennia before the Common or Christian Era. This calls out for explanation. Explanations have indeed been provided, and they are, as I have suggested, revealing of changes in modern British culture. The one that was dominant during my own adolescence, and throughout the mid-twentieth century, had many virtues. One of these was simply unanimity: it was accepted by all the leading experts in British prehistory, in alliance with colleagues who specialised in all other parts of Europe. Another virtue was its longevity and consistency: that it had been developing for one and a half centuries before reaching its apogee in the 1950s and early 1960s. It portrayed the tomb-shrines as having been the temples of a religion dedicated to a single Great Goddess, who represented the earth and the generative powers of nature. It held that this had first arisen in the Near East, and been brought to Western Europe by missionaries, until it covered the entire continent and the whole Mediterranean basin. The tomb-shrines were often thought to represent her body, in which the dead were laid to await rebirth, and in which they acted as mediators between the human and divine worlds. In this traditional view, the Goddess's worship was brought to an end by invaders from the steppe country which bordered Eastern Europe. These introduced a new religion, focused on the sky and above all on the sun, and on the element of fire which was associated with it. It was they who established the new circular temples, mirroring the solar orb, the new round burial mounds, and the rite of cremation, by which the dead were committed to the sacred fire.[8]

The newcomers also brought a fire-based technology consisting of metalworking, in gold, copper and bronze, replacing the stone tools and weapons of the tomb-shrine era. This gave them the military superiority which enabled them to conquer and absorb the tomb-shrine builders. They

[7] For all this see Hutton, *Pagan Religions*, pp. 52-87, and subsequently Hutton, *Pagan Britain,* Chapter 3.

[8] Ronald Hutton, 'The Neolithic Great Goddess', *Antiquity* 71 (1997): pp. 91-99.

were, moreover, of a different race to those whom they subdued, being taller, blonder and with blue eyes, whereas the Stone Age folk of the tomb-shrines were small and dark. They therefore had the edge in physical as well as technological prowess. The sense of who these invaders were changed over the course of the twentieth century. In the late Victorian period they were regarded as the Celts, but over the next generation these were shifted to the Iron Age, as the last great wave of prehistoric newcomers to Britain. Instead the bringers of the solar religion of circles and fire became the Indo-Europeans, given the specific form in Western Europe of the Beaker People. These were named after the distinctive drinking vessels found in graves beneath the early round barrows. The beakers were one component of a complete assemblage of newly appeared weapons, tools and ornaments which archaeologists interpreted as the trappings of a warrior society.[9]

This model of change was deeply satisfying to a range of personality types and interest groups in modern British (and European) society. For one thing, it provided a dramatic and lucid story that appeared to fit the archaeological evidence. For another, it could be retold with a number of different infusions of sympathy. For those emerging into a post-Christian society, and experiencing a need to engage imaginatively with the divine feminine, the concept of a primordial Great Goddess was deeply attractive. It could be given a deeper feminist hue by suggesting that the small, dark people of the Neolithic, who worshipped her, also had a woman-centred society, more pacific and ecologically friendly than those after it. This had the effect of making the arrival of the Beaker People all the more tragic, as it could be made to represent not only the replacement of a matriarchal with a patriarchal religion, but an equivalent change in society. In this view, it took the form of the destruction of a peaceful, consensual, responsible and feminist order by violent, patriarchal brutes, who introduced a system based on inequality and exploitation, which glorified war and masculinity and had a polluting and extractive technology. The converse interpretation was to glorify the coming of the Beaker People and the solar religion as a great forward step in the progress of humanity. This

---

[9] See for example, Jacquetta Hawkes, *Early Britain* (London: Collins, 1945), pp. 18-23; Jacquetta Hawkes and Christopher Hawkes, *Prehistoric Britain* (Harmondsworth: Penguin, 2nd edition, 1949), pp. 66-71; Stuart Piggott, *The Neolithic Cultures of the British Isles* (Cambridge: Cambridge University Press, 1954), p. 270; D. L. Clarke, *Beaker Pottery of Great Britain and Ireland* (Cambridge: Cambridge University Press, 1970), pp. 276-80.

characterised the Neolithic, with or without a woman-centred society, as having been more ignorant and savage than the succeeding Bronze Age. It glorified, or at least respected, the Beaker People as bringing a more sophisticated society, as well as a much more superior technology. Their arrival, in this vision, was one of the first great steps taken by European humanity in its long march towards the benefits of modernity. It is not difficult to see that, between them, these two different approaches to the same basic story summed up the two opposing attitudes of modern Westerners to their age and to their society.

Both of the components of the story—the Great Goddess and the Beaker People—had deep roots, and had converged from separate points of origin. The concept of a universal goddess, identified with the natural world, drew upon ancient ideas but had become dominant in the Western literary imagination with the coming of the Romantic Movement. As such, it was explored by poets and novelists all through the nineteenth century, and in 1849 it was back-projected by a German classicist, Eduard Gerhard, into the ancient past. He became the first scholar to propose that such a goddess had been worshipped by all the peoples of the prehistoric Mediterranean and Near Eastern worlds. In his reading, her figure had subsequently fragmented into the many goddesses and gods found in actual ancient pantheons when history began. This idea was gradually taken up by other German and French scholars in the rest of the century, and adopted by their Britain colleagues in the early twentieth. Subsequent archaeological discoveries were promptly interpreted in harmony with it, creating a larger and larger structure of apparent evidence.[10] The concept of invasions as the motor for prehistoric change was also a development of the mid-nineteenth century, which spread from the Continent to Britain. This time it was the Danes who proposed it, in the 1840s, but its adoption was far more rapid than the idea of the Great Goddess. It was an orthodoxy of British scholarship by the 1860s, and fully elaborated by the 1880s. The reason for its instant appeal is very clear. What it outlined was a story of British prehistory in which technological and social change was introduced by successive waves of aggressive newcomers, each more advanced than the last.[11]

---

[10] Ronald Hutton, 'The Discovery of the Modern Goddess', in Joanne Pearson, Richard H. Roberts and Geoffrey Samuel (eds), *Nature Religion Today* (Edinburgh: Edinburgh University Press, 1998), pp. 89-100.

[11] Ronald Hutton, *Blood and Mistletoe: The History of the Druids in Britain* (London: Yale University Press, 2009), pp. 299-303.

What clearly underlay this picture was the reality of European imperialism in the same period, during which European or Europe-derived states were spreading their rule across ever-larger areas of America, Africa, Asia and Australia. English-speaking nations in particular were sending large bodies of settlers into these lands, which were dispossessing and sometimes destroying the native peoples who had occupied them. This experience explicitly underlay the developing Victorian view of British prehistory. An explicitly racist element was injected into this view by the belief that each successive wave of incomers had been taller, blonder and stronger than the last. The greatest single replacement had been of the small dark people of the Neolithic, with their preoccupation with earth, by the tall fair people of the Bronze Age, with their sights set on the sky. The argument that remnants of the older, inferior, race were still found among the modern population of the islands enabled the Victorian British elite to present pseudo-scientific reasons for despising particular subsets of it. The Irish were the main victims, but sections of the British working class were also targeted. The scholar responsible for the idea of this genetic replacement was a medical doctor, John Thurnham, who made two major contributions to the study of British prehistory. One was to point out that the tomb-shrines and the round barrows actually belonged to different millennia, instead of, as assumed hitherto, being built by the same people. This was a real, and permanent, advance in knowledge. He also, however, added the assertion that they were made by different races. This was unsupported even by his own data.[12] The subsequent addition to his ideas, that the tomb-shrine people were small and dark and the stone circle people tall and blonde, was absurd. Their skeletons are actually of the same size, and you can't tell a person's complexion from bones. None the less, these Victorian beliefs lasted until the mid twentieth century, largely because so many other Victorian structures did: empire, great power status, racism, gender polarity, and an economic dependence on heavy industry, reliant on extracted minerals. A further component in the invasion model also endured: the notion of Britain as an island threatened by foreign attack. This repeatedly surfaces in accounts of its prehistory, and actually strengthened through the early twentieth century, because of two World Wars and then the beginning of the Cold War.[13]

---

[12] John Thurnam, 'On Ancient British Barrows', *Archaeologia* 42 (1869): pp. 161-244; and *Archaeologia* 43 (1871): pp. 285-544.

[13] For examples, see Hawkes and Hawkes, *Prehistoric Britain*, p. 13.

The whole traditional vision of Neolithic and Bronze Age British prehistory unravelled during the 1960s and 1970s, and by 1980 it was gone, although it still has echoes in popular works, both of fiction and non-fiction, to this day. I was myself a witness of the whole process of disintegration, at close quarters, and so I can speak about both the public and the largely unspoken factors involved in it. The greatest was simply that the two decades concerned witnessed the end of Victorian Britain, in all the aspects described above: the empire, Great Power status, a fear of invasion by land forces (as opposed to missiles), an economic dependence on heavy industry, and an official tolerance of racism and sexism. A Victorian model of prehistory now became vulnerable. Another factor working for change was the great expansion of higher education in the same period, creating many new experts in prehistory. Linked to the general disrespect for traditional ideas which was also a feature of the age, this prompted a wholesale questioning of received models.

The final major development relevant to our subject was the improvement in dating techniques for ancient sites, based on the analysis of radiocarbon, which became available around 1970. Combined with improved statistical analyses of data, these have permitted more and more precise dates to be achieved for prehistoric material. In itself, this single scientific innovation rendered the old model untenable. It shattered the presumed chain of transmission for the religion of the Great Goddess from the Near East to Western Europe. Much of the evidence provided for it in the Mediterranean basin turned out to be younger, not older, than that on the Atlantic seaboard. The dating revolution was even more lethal to the idea of Beaker People invasions. All of the innovations that had been associated with those—circular monuments, cremation, metalworking, and the range of specific prestige goods—were proved to have arrived, slowly, at different times in the period between 3200 and 2200 BCE. They were not part of a single cultural package.[14] Advances in genetics, especially in the analysis of DNA, proved that there had been no significant arrival of a new racial group in the whole of the period concerned.[15] A much enlarged body of excavated material showed that the earlier Neolithic, the time of the tomb-shrines, was actually more warlike than the later Neolithic and

[14] Hutton, *Pagan Religions*, pp. 16-138.

[15] Stephen Oppenheimer, *The Origins of the British* (London: Constable and Robinson, 2006), pp. 210-370; Brian Sykes, *Blood of the Isles* (London: Bantam, 2006), passim.

Bronze Age, the time of the stone circles.[16] The question of whether women or men led prehistoric societies has become pointless because we completely lack any decisive evidence for the nature of social structures in earlier British prehistory. From the same data, you can visualise matriarchy, patriarchy, theocracy, democracy or tribal chieftainship, as you please.

There is no better evidence for the nature of the deities worshipped in the British Neolithic or Bronze Age: you can imagine what you want. The concept of a universal Great Goddess was abandoned by all British, and most European and American, archaeologists because there was nothing solid to sustain it. It remains possible, and it must be emphasised that it is pretty well certain that the prehistoric British believed in goddesses, or at least in powerful female spirits. Traditional peoples always do. This is, however, a very different thing from believing in one single all-powerful deity, associated with the earth, across the entire ancient Eurasian world: that looks very much like a modern construct. Only towards the very end of the ancient pagan world did such monist or monotheistic religious ideas begin to be articulated, and these were never embraced by the majority of pagans even then.[17] To my own generation of scholars the idea of the Neolithic Great Goddess had three features which made it especially unappealing. The first was that it seemed so clearly a post-Christian construct: of a single, universal, primeval religion, of a single deity, which later degenerated into the polytheism of the historic ancient world. The second was that it seemed to present such an essentialist concept of femininity: of the female as mother, nurturer, representative of fertility and regeneration. Many of the historic pagan goddesses, as patronesses of rulership, science, crafts and wisdom, seemed much more attractive as role models for modern feminism. The third drawback was that it embodied a sharply polarised view of masculine and feminine. Anthropology was now furnishing us with huge quantities of new information about the ways in which gender relations had been constructed in non-European societies. It showed us the great range of possibilities which were actually open to us. By 1975 one British anthropologist, Shirley Ardener, could pose the

---

[16] A sample of a large literature on this is Roger Mercer, 'The Origin of Warfare in the British Isles', in John Carman and Anthony Harding (eds), *Ancient Warfare* (Stroud: Sutton, 1999), pp. 143-56.

[17] See for example, Polymnia Athanassiadi and Michael Frede (eds), *Pagan Monotheism in Late Antiquity* (Oxford: Oxford University Press, 1999).

exciting question of whether our Western categories of 'woman' and 'man' might not disappear altogether.[18]

To many British prehistorians, therefore, it was a shock when the old ideas came back to Britain from America in the 1980s, but this time as part of radical feminism. American writers had taken up the old idea of an essential female nature and simply attached a positive value to those aspects of it which had often been treated as negative. This movement attracted the support of one distinguished archaeologist, Marija Gimbutas, the leading Western expert in eastern European prehistory. She reasserted the whole traditional idea of a Goddess-centred, pacific and creative Neolithic Europe, destroyed by Indo-European warriors worshipping sky-gods. She simply gave it a new liberationist message.[19] In other words, British and American radicals had dealt with the shortcomings of the old model in opposite ways. The former had deconstructed it; the latter had appropriated and reshaped it. Both are excellent strategies for dealing with an inconvenient intellectual construction. The problem is that they are completely mutually incompatible. As a result, the very British academics who had supported the demolition of the Great Goddess construct in the name of socialism, feminism and gay liberation now found themselves being abused as patriarchs and reactionaries by followers of the new American Goddess movement. My concern here, however, is with what those same British academics put in place of the Goddess and the patriarchal invaders. What they provided, in brief, was Marxism, the most dynamic intellectual movement in the years around 1970 in which many of them were educated. In one aspect, this produced a secularisation of prehistory, depriving religious belief of any status as a force in itself and grounding all ideology ultimately in economic needs and the power politics that they generated. The tomb shrines were therefore now interpreted as territorial markers, built by people who were taking on the new Neolithic farming lifestyle. This involved settling down on the land and dividing it up, and the new monuments served to warn strangers that particular plots were already taken. The human bones inside them were interpreted as those of the first people to occupy that farm, who were then revered as ancestors by their successors. This was part of a continuing process of affirmation of group identity and rights of possession. The transition to the age of the circles was seen as marking a shift from a society based mainly on those group identities to one in which individuals were more prominent.

---

[18] Shirley Ardener (ed.), *Perceiving Women* (London: Malaby, 1975), p. xviii.

[19] Hutton, 'The Neolithic Great Goddess', pp. 97-98.

In the tomb-shrines the bones had been mixed together in large monuments requiring considerable collective effort. In the round barrows people were buried individually, and the most important had personal possessions interred with them. This model certainly seemed to explain why the British in the third and second millennia BCE were apparently so fond of consumer goods—weapons, tools, pottery and ornaments—in increasing numbers and variety. With equal certainty, it suited British society in the 1970s and 1980s, both in its secularism and in its stress on the individual. After all, the British in the mid-twentieth century had themselves passed from modes of behaviour which had largely been based on collective and conformist models to a rampant individualism based largely on new and rapidly-changing fashion accessories.[20]

The Marxist system of explanation, however, always left major parts of the evidence unexplained. One was the new interest in the circle as the vital unit of sacred space. Another was why people moving towards the new individualism should still engage in huge collective building works such as Stonehenge, Avebury, Maes Howe and Newgrange, which dwarfed that needed for the tomb-shrines. It foundered completely when more was discovered about the early Neolithic way of life. This was not in fact based on an agrarian economy of farms and fields, but on a pastoral one of people migrating with flocks and herds along seasonal routes. The clusters of tomb shrines could not, therefore, have marked out family plots. Furthermore, the bones in them were added at successive intervals, and so could not have commemorated founding ancestors. They do seem to indicate a religion mediated at least partly through the dead.[21] The model of British prehistory based on religion and race had developed and flourished for a hundred and fifty years. The Marxist one founded after less than thirty. In the twenty-first century, none has appeared to take the place of either. Instead we have a range of individual suggestions from different experts. One is that the very process of constructing huge monuments, needing as it did project leaders, helped to create a new elite class of individuals.[22] Another is that a sense of the sacred which was traditionally focused on places, and hence on monuments, became refocused on

---

[20] E.g., Richard Bradley, *The Social Foundations of Prehistoric Britain* (London: Longman, 1984); Julian Thomas, 'Reading the Body', in P. Garwood et al. (eds), *Sacred and Profane* (Oxford: Oxford University Press, 1991), pp. 33-42.

[21] For this see Hutton, *Pagan Britain*, Chapters 2-3.

[22] John Barrett, *Fragments from Antiquity* (Oxford: Blackwell, 1994), pp. 27-32.

humanity, and so on prestige goods.[23] A third is that people passed from honouring multiple ancestors to a single ancestor, so that the dead were remembered by their goods rather than by their bones.[24]

It must be obvious that none of these explains the change in the form of monuments—from tomb-shrines to circles—around 3000 BCE. A couple of other recent analyses have acknowledged the important point that a shift to single burials, with prestige goods, under round mounds was not just a British phenomenon but a Europe-wide one in the third millennium BCE. It seems to have spread from east to west across the continent, and Richard Harrison and Volker Heyd, of the Bristol Archaeology Department, have credited it to a new ideology. This emphasised material objects as the basis for personal identity and social position, and venerated the sun as the focus of religion.[25] Another prominent British archaeologist, Timothy Darvill, has also found evidence for enhanced sun worship in the new interest in circles and orientation of monuments on solstices. He has identified solar imagery in designs on stones and pottery at the same time.[26] Some place is therefore now being made again for religious factors in analyses of the changes around 3000 BCE, but only by a minority among experts. This, and the lack of any prevailing theory of explanation for the changes concerned, clearly suits our contemporary social world, of a dominant secularism and a celebration of individualism and diversity. I would suggest that it is both important and necessary to note what is missing in it.

Race is obviously gone as an explanatory force, for perfectly obvious and good reasons, but so have invasion and migration. Instead the new fashions which spread across Europe, and took such dramatic forms in Britain, are credited to individuals, who arrived as salespeople, traders, marital partners and migrant workers, bringing the relevant fresh ideas and technologies. This is, of course, a perfect projection, onto prehistory, of the world of the current European Union and the global economic order. It certainly can fit the archaeological and genetic evidence: but there is a problem with it. This is that invasions, and migrations of ethnic groups, are

---

[23] Jan Harding, *Henge Monuments of the British Isles* (Stroud: Tempus, 2003), pp. 112-21.

[24] Andrew Jones, 'How the Dead Live', in Joshua Pollard (ed.), *Prehistoric Britain* (Oxford: Blackwell, 2008), pp. 177-201.

[25] Richard Harrison and Volker Heyd, *The Transformation of Europe in the Third Millennium BC* (Berlin: De Gruyter, 2008), passim.

[26] Timothy Darvill, *Prehistoric Britain* (London: Routledge, 2nd edition, 2010), pp. 132-200.

a major theme of recorded ancient history. As soon as Britain emerges into the historical record, parts of it were occupied successively by Roman, Anglo-Saxons, Irish, Vikings and Normans. The Roman Republic was at times attacked and endangered by warlike people from the north, and the Western Roman Empire of course succumbed to them. Phoenicians and Greeks established maritime colonies across the Mediterranean. Further back, in the second millennium BCE, one group of invaders, the Hyksos, brought down the Middle Kingdom of Egypt, and another, the Sea Peoples, fatally weakened the succeeding Egyptian New Empire. In the same period waves of predatory incomers—Amorites, Kassites and Aramenaeans—destroyed successive states and civilisations in Mesopotamia. Experts in British prehistory, however, will not admit to one single significant military incursion or migration into Britain in the whole of the last four millennia BCE. The DNA evidence may actually not be very helpful here, if people on both sides of the North Sea and English Channel already had quite similar genes by the Neolithic. Perhaps everything did change dramatically, and all hell did break out as soon as history began, but if it did then such a remarkable phenomenon deserves more discussion than it is receiving. After the collapse of Marxist scholarship, ideology is now once more being given more recognition than before as a force in its own right, but there is still reluctance among most British archaeologists to accord this to religious ideology. Even when explanations are permitted in religious terms, they tend to be in the form of heavenly bodies—for example 'the sun'—rather than in terms of the deities to whom such bodies are commonly related in traditional societies. The study of the British Neolithic and Early Bronze Age matters, because this period produced some of the most spectacular prehistoric monuments on earth. They include between them three World Heritage Sites. It is currently involving more specialist scholars, with more students and a more sophisticated range of technological and intellectual aids at their disposal than ever before. It would be both impudent and reckless of me, therefore, to suggest that it is currently largely neglecting no less than three of the most important areas of human experience. They are the sky, on which I have offered a paper in a previous volume of the present series; invasions and migrations; and the worship of goddesses and gods, or at least of potent spiritual beings.[27]

---

[27] Ronald Hutton, 'The Strange History of British Archaeoastronomy'.

Galileo and the Astrological Prophecy of Manuel Rosales

Nick Kollerstrom

Abstract: Scholars and biographers of Galileo have felt at liberty over the centuries to ignore his apparently keen endorsement of a polemical and astrological prophetic text, which he published in Rome in 1626. Omitted from the collected works of Galileo, it was finally brought to light by Luigi Guerrini in 2001. The text concerned a Portuguese prophecy for the restoration of its empire. Such delayed recognition may serve to remind us of the sheer extent to which Galileo's involvement in astrology has been censored, marginalised and written out of the history books. The prophecy was viewed as inflammatory by the Spanish authorities, and orders went forth for all copies to be destroyed. One single copy remains of this publication with Galileo's foreword, in the *Biblioteca Nazionale Central* in Florence, where it was found by M. Guerrini. The author's name seems to have varied over the years from Manuel Bocarro Francês y Rosales to Dr Jacob Rosales to Imanuel Bocarro Francês. It is unclear whether Galileo and Rosales ever met; it is possible that Galileo wrote this Foreword at the request of the Medicis who employed him.

> Dr Jacob Rosales, [was] a many-sided, exceptionally gifted and controversial personality, a prolific author interested in astronomy, mathematics, medicine, alchemy, literature, politics, political astrology and Jewish apologetics and one of the best-known exponents of political messianism, whose alchemy and astronomy studies led him onto prognostication. He was also an accomplished poet writing in at least three languages, Portuguese, Spanish and Latin.[1]

The Prophecy

In 1624 the Jewish philosopher Manuel Bocarro Frances y Rosales, described by Halevy and Silva as 'a prolific author interested in astronomy,

1 M. Halevy and S. Silva, 'Tortured memories, Jacob Rosales alias Imanuel Bocarro Frances: a life from the Files of the Inquisition', *The Roman Inquisition, the Index and the Jews*, ed. S. Wendehorst (Leiden: Brill, 2004), pp. 107-51, p. 117.

Nick Kollerstrom, 'Galileo and the Astrological Prophecy of Manuel Rosales', *Culture and Cosmos*, Vol. 18 no 1 (Spring/Summer 2014), pp. 35-41.
www.CultureAndCosmos.org

mathematics, medicine, alchemy, literature, politics, political astrology and Jewish apologetics and one of the best-known exponents of political messianism', published his prophecy, composed with a view to the restoration of the Portuguese monarchy, featuring some millennial-apocalyptic visions.[2] Entitled, 'Small lunar and smaller light of the Portuguese monarchy: Explanation of the first *Anacephaleoses*', it was printed in Lisbon. It foresaw a 'hidden Prince' who would restore the monarchy and specified the year 1653 when he would appear, to rule the world.[3] One part of this book was entitled *Luz Pequena*, a development of the four-part astrological poem published in 1624, *Anacephalaeoses da Monarchia Luzitania,* 'A Summary of the Lusitanian Monarchy'. It was a prophetic and messianic work of which only part one seems to have actually appeared, the other three being burnt immediately by the Spanish authorities. Utilising methods of political astrology, the book predicted that Portugal, then under Spanish rule, was destined to become the last and mightiest world empire, when in the year 1653 the emergence of the 'hidden king' would vanquish the followers of Mohammed.

With Portugal under Spanish rule, copies of this seditious work had to be burnt by the authorities, and its author Bocarro-Rosales denounced for a second time before the Holy Inquisition. This was done by his own brother in 1624 at the Goa Inquisition Court (Goa was then a Portuguese Indian colony); then Mr Bocarro was denounced one more time in 1626 at Lisbon. It was a stressful time for Jews in Europe and altogether he was denounced to the Holy Inquisition on nine occasions before the courts of Goa, Lisbon and Madrid. These Inquisition reports contain most of the known information about his life.

Upon being released from jail, Bocarro fled to Rome, changed his name and 'came out' as a professing Jew (as opposed to a Marrano Jew— one obliged to convert to the Christian religion, as he had to while in Spain). He arrived in Rome in 1625; Galileo would not have been there, as he was then living near Florence. Guerrini described how, around the middle of 1625, Mr Bocarro arrived 'in Rome, at Gomez e Silva, Duke of Pastrana, a character very influential and very close to the dominant Barberini family'.[4] He seems to have adopted the name 'Rosales' upon arriving in

---

[2] Halevy and Silva, 'Tortured memories', p. 127.

[3] Halevy and Silva, 'Tortured memories', p. 126.

[4] Luigi Guerrini, 'Galileo fra gli astrologi', *Bruniana e Campanelliana* 7, no.1 (2001): pp. 7, 233-44, 238; republished in *Ricerchesu Galileo e il primo Seicento* (2004): pp. 97-105, p. 100.

Italy. There were six occasions on which Galileo visited Rome: the first was in 1616, and then later from April to June of 1624 to celebrate the election of the new pope Barberini.

What Galileo Saw

Galileo assisted the publication, in Rome, of this astrological prophecy. The part of this book prefaced by Galileo was entitled *Luz Pequena*. As published in Rome in 1626, Rosales' *Luz Pequena* comprised 30 pages in Portuguese, in both prose and verse, with Galileo's introduction in Latin. It was the fourth part of his banned book *Anacephalaeoses*, together with notes which he brought out under the title *Luz Pequena Lunar*, the first three parts having appeared in 1624.[5] Galileo may have translated some parts of it. It alluded to the Copernican theory and the eccentricity of the solar orbit, but hardly in a style one might expect Galileo to have approved of.

It seems not to have troubled Galileo that the work had been banned in Portugal where it was printed, and its Jewish author thrown in jail. His introductory paragraph praised the 'learned astrologer' who had composed it: Manuel Rosales, he explained, had 'sent us as a gift his astrological judgements, similar to prophecies, in his excellent Portuguese language', adding, 'We have taken care to translate it into the Italian language'.[6] (If Galileo had produced such a translation, no copies remain.[7]) Galileo's introduction praised this text as being 'in the author's own words, since they are of great meaning for general knowledge and love of science' and described Rosales as 'first amongst astrologers'.

His Introduction to Rosales' text alluded to 'this very admired and extremely learned Doctor Emmanuel Bocarro Frances (*Virum Admirandum, & doctissimum Astrologorum Principem*) 'who also rejoices in the name Rosales' (Latin '*gaudet*' for 'rejoices in'), which implies—or so Moreno-Carvalho has surmised—that 'the letter was the outcome of a personal meeting between Rosales and Galileo' where Galileo had the

[5] Halevy and Silva, 'Tortured memories', p. 128; Luis M. Carolino, 'Scienza, politica ed escatologia nella formazione dello "scienziato" nell'Europa del XVII secolo: il caso di Manuel Bocarro Francês - Jacob Rosales', *Nuncius* 19, no. 2 (2004): pp. 477-506.

[6] Text and translation given at end of essay.

[7] Its accession number in Florence national Library is 1068.27. The text was edited in 2006 by Luis Miguel Carolino, and the Biblioteca Nacional de Portugal owns a copy.

significance of this name-change explained to him.[8] He adopted the name Rosales upon arriving in Rome. Or, does Galileo's preface suggest he had been *sent* the document? As in, Mr Bocarro 'sent us as a gift his astrological judgements, similar to prophecies'.[9] But, the latin '*obtulerit'* in Galileo's Preface may not be optimally translated as 'sent', but could rather imply that he had received Bocarro's booklet from somebody who was Portuguese.[10] Galileo then added the words of great praise: 'So, I invite him who explains the book of the universe to admire, love, and praise the talent of this man'.[11]

It is hard to think of anyone except Giovanni Sagredo who gets such an enthusiastic and affectionate write-up from Galileo. If these two had not met, then the affectionate and familiar tone of this Preface is quite hard to explain. He signed it simply as 'G.G. Mathem'. But no-one has doubted that this was really Galileo. As court philosopher to the Medici family in Florence, Galileo had described himself in his 1624 *Il Sagiattore* as a 'philosopher', that is, no longer just a 'mathematicus' as he had been in Padua. He may well have been asked by the Medicis to get this text published: the Spanish crown was seen as a threat by the Vatican, so it could have been politically advantageous to have their court philosopher Galileo endorse a prophecy of its downfall; although Galileo himself generally kept clear of politics, this could have been important for the Medici family. Support for this view comes from a much later work of Rosales published in Florence and dedicated to Cosimo III de Medici.[12]

This text has been omitted from just about all collections, anthologies, indexes of letters, and books about Galileo, over the centuries—certainly all English-language texts—until finally Luigi Guerrini noticed it, and

---

[8] F. Moreno-Carvalho, 'A Newly-discovered letter by Galileo Galilei: contacts between Galileo and Jacob Rosales (Manoel Bocarro Francês), a seventeenth-century Jewish scientist and Sebastienist', *Aleph* II (2002): pp. 59-91; he surmised that Rosales may have met Galileo in Florence, see p. 78.

[9] The translation given by Moreno-Carvalho, 'Galileo and Jacob Rosales', p. 73.

[10] Advice from Luigi Guerrini, who supplied the translation used here.

[11] See text translation at end of essay.

[12] Rosales' 1754 work *Fasciculus Trium Verarum Propositionum Astronomicae, Astrologicae at Philosophicae* was published in Florence and dedicated to Cosimo III de Medici; see Carolino, 'Scienza, politica', p. 482.

published his report on it in 2001. A publication of 2002 by Moreno-Carvalho wrongly claimed to have been the first to publish it.[13]

A Time of Optimism

Jupiter-Saturn conjunctions take place every twenty years, and their sequence remains in a given zodiac element or 'trigon' (a triangle of three zodiac signs with the same element) for two centuries, so that it moves through all four zodiac-elements every eight hundred years. These conjunctions were traditionally accepted as measuring out the cycles of history and an especial importance was attributed to their entry into the fire-element, as happened at the dawn of the 17th century with the 1603 triple conjunction. Based on this, Tycho Brahe had envisioned the return of a Golden Age. Tycho Brahe's vision of 'the start of a new golden age' was based partly on the new star he had seen, in 1572, but also on his view that 'the seventh cycle of history was about to begin', counting these by the chronocrators (Jupiter-Saturn conjunctions) moving through the zodiacal elements.[14] On the other hand, Kepler in his *De Stella Nova* of 1606 had been more circumspect, though pointing out that the previous entry into the fire-element was in the time of Charlemagne and the one before that, the birth of Christ; there had been 'seven great periods' of 800 years since the 'beginning of the world' 5,600 years ago.[15] The visionary-astrologer and political revolutionary Tomasso Campanella likewise prophesied the 'return' of the Golden Age.[16] Campanella followed Tycho Brahe, and his argument was in some degree based upon the notion of history as having as periods of six thousand year duration, as well as the two new stars appearing in the sky. In his very first public lectures on astronomy made in 1604 about the bright new star which had appeared in the sky, Galileo

---

[13] Darrell Rutkin's 'Galileo Astrologer: Astrology and Mathematical practice in the late-sixteenth and early seventeenth centuries', *Galileana* II (2005): pp. 107-143, gives Guerrini credit for this discovery, see p. 136.

[14] Germana Ernst, 'From the Watery Trigon to the Fiery Trigon: Celestial signs, Prophecies and History', in *Astrologi hallucinati' Stars and the end of the World in Luther's Time*, ed. Paola Zambelli (NY: Walter de Gruyter, 1986), pp. 265-80, p. 272.

[15] For English translation of Chapters 7-9 of Kepler's *De Stella Nova* (Frankfurt, 1606), see 'Kepler's Astrology', *Culture and Cosmos* (Winter 2010): pp. 209-34, p. 213.

[16] Germana Ernst, 'Watery Trigon', p. 270; also Germana Ernst, 'Astrology and Prophecy in Campanella and Galileo', in *Galileo's Astrology*, ed. Nicholas Campion and Nick Kollerstrom, *Culture and Cosmos* 7, no. 1 (2003): pp. 21-36.

discussed how it had been right next to that Jupiter-Saturn conjunction of 1603 in the sign of Sagittarius: intimating that the new star had somehow been produced by the conjunction. That had been a rare once-per-century triple conjunction, with Jupiter thrice crossing the longitude of Saturn, i.e. meeting Saturn three times.[17]

Rosales' prophecy, endorsed by Galileo, was based upon the appearance of the very bright comet of 1618, and upon the last two Jupiter-Saturn conjunctions. Campanella saw Galileo's own great discoveries as fulfilling this prophecy. There are nine letters which Campanella wrote to Galileo which survive, some quite lengthy, but no replies extant from Galileo.[18] We may, I suggest, best apprehend whatever side of Galileo's character impelled him to write in praise of Bocarro, by trying to evaluate his correspondence with Campanella. The latter explained to him that the prophecy of a new heaven and a new earth was being fulfilled by Galileo: after all, had not his telescope seen the new heaven?[19]

In 1644, Rosales republished his prophecies, including Galileo's Foreword. He felt vindicated because a Portuguese monarchy 'was indeed restored in the fateful year of 1640'.[20] A text of his claiming his prophecy had been valid was published in Florence in 1654. All copies of Bocarro's work had to be burnt for their 'heresy' according to an edict of 1774, but maybe this decree did not reach Florence, where a copy remains in the main library.[21]

---

[17] Campion and Kollerstrom, *Galileo's Astrology,* pp. 73-74.

[18] Germana Ernst, *Tommaso Campanella: The Book and the Body of Nature* (New York: Springer, 2010) p. 159.

[19] Germana Ernst, 'From the watery Trigon to the fiery Trigon: Celestial Signs, Prophecies and History', in Zambelli, Paola (ed.), *'Astrologi hallucinati': Stars and the End of the World in Luther's Time',* Berlin and New York, Walter de Gruyter 1986, pp. 265-80 (pp. 265-6); also, Germana Ernst, *Astrology and Prophecy in Campanella and Galileo* (English translation of Ernst's essay in *Novita celesti e crisi del sapere*, ed. P. Galluzi, Florence 1983), in Campion and Kollerstrom, *Galileo's Astrology*, p. 24; see also Germana Ernst, 'Galileo, Campanella e le dottrine celesti', in *Il processo a Galileo Galilei e la questione galileiana*, ed. G. M. Bravo, *Bruniana & Campanelliana* 15, no. 1 (2010): pp. 159-84.

[20] L. M. Carolino and C. Z. Camenietzki, 'Tokens of the Future: comets, astrology and politics in early modern Portugal', *Cronos* 9 (2006): pp. 33-57, p. 34.

[21] Moreno-Carvalho found one other copy of the *Luz pequena,* in the general library of the Coimbra university in Portugal: 'Galileo', ref. 1 p. 72, entitled*: Luz Pequena lunar et estellifera. Do Doutor Manoel bocarro Francez Rosalez.*

Text:
Lectori amico. Hoc viri admirandi, et supra modum doctissimi DMBF, qui etiam R nomine gaudet, iudicium astrologicum, vaticinio simile, ad nostra pervenit manus, cum excell. Personae, lusitano idiomate, illud obtulerit. Et quamvis huius modi opusculum cum I Anacephal. De quo agitur, converti in italicum sermonem curavissemus, sic quae eo fruamur, nihilominus, typis mandare propria autoris verba, sunt enim magis significativa, ob commune studium et scientiae amorem curavimus, ut ad hibito, quem exponit, libro mundus viri astrologorum principis, ingenium miretur, amet et laudet. Romae 1 Julii anno 1626 G.G. Mathem.

Translation by Luigi Guerrini.[22]
To the friendly reader, this astrological judgment, in the manner of prophecy, of the admired and extremely learned doctor I. Manuel Bocarro Frances, who is also called Rosales, came through our hands. Since it is written in the Portuguese language of its distinguished author, we took care to translate this booklet, together with I Anacephal (which deals with it) into Italian, thereby enjoying it. Nonetheless, for common study and love of knowledge, we decided to send into print the author's own words, which are in fact most significant, so that people could admire, love and praise the talents of this man, first among the astrologers.
Rome, July 1st 1626, G. G. Mathem.

[22] Translation kindly provided by Luigi Guerrini for the author.

Dr Katterfelto and the Prehistory of Astronomical Ballooning[1]

Clive Davenhall

Abstract: Regular telescopic astronomical observations made from balloons began after World War II, though scientific, particularly meteorological, ballooning dates from the mid-nineteenth century. However, astronomical ballooning has a curious prehistory at the dawn of lighter-than-air travel in the 1780s. The self-styled Dr Katterfelto (c.1743?-99) was a German-born travelling showman, lecturer and considerable self-publicist who in 1784-85 claimed to have made important astronomical discoveries from observations made from a balloon. It is unlikely that he made any such observations, or, indeed, any balloon flights. However, the episode throws some light on the world of the itinerant, eighteenth-century astronomical lecturer and the diffusion of contemporary astronomical and scientific knowledge.

Introduction

Gustavus Katterfelto (c.1743?-1799; see Fig. 1) was a German-born travelling showman and lecturer who performed throughout England and Scotland from 1776 until his death in 1799.[2] In Britain during the eighteenth century there was a tradition of itinerant lecturers who gave talks and demonstrations on scientific and other subjects (see Fig. 2).[3]

---

[1] This paper was presented at the INSAPVII conference in Bath, October 2010.

[2] D. Paton-Williams, *Katterfelto: Prince of Puff* (Leicester: Matador, 2008); see also P. Fara, 'Katterfelto, Gustavus', in *Oxford Dictionary of National Biography* (Oxford: Oxford University Press, September 2004); online edn., updated October 2007, at http://www.oxforddnb.com/view/article/15187 (accessed 1 October 2010).

[3] For an overview see H. C. King in collaboration with J. R. Millburn, *Geared to the Stars: the Evolution of Planetariums, Orreries and Astronomical Clocks* (Toronto: University of Toronto Press, 1978), Chapter 9. For more detailed discussions in the limited context of travelling lecturers who worked in Bath see T. Fawcett, 'Science Lecturing at Bath 1724–1800', *Bath History*, Vol. vii (Bath: Millstream Books, 1998), pp. 55-57; or T. Fawcett, 'Science Lecturing in Georgian Bath', Chapter 11 of P. Wallis, ed., *Innovation and Discovery: Bath and*

Clive Davenhall, 'Dr Katterfelto and the Prehistory of Astronomical Ballooning', *Culture and Cosmos*, Vol. 18 no 1 (Spring/Summer 2014), pp. 43-51.
www.CultureAndCosmos.org

Fig. 1: Gustavus Katterfelto (c.1743?-1799). On the right, Katterfelto's solar-illuminated microscope is being used to show bacteria. This woodcut appeared in the *The European Magazine* and *London Review* for June 1783, p. 406.

The lecturer and author James Ferguson (1710-1776) is an example of a well-respected practitioner in this tradition, whose lectures and demonstrations had a serious intent.[4] Katterfelto, by contrast, was a

the Rise of Science (Bath: Bath Royal Literary and Scientific Institution and The William Herschel Society, 2008), pp. 144-51.

[4] J. R. Millburn in collaboration with H. C. King, *Wheelwright of the Heavens: the Life and Work of James Ferguson* (London: Vade-Mecum Press, 1988); P. Rothman, 'Ferguson, James (1710–1776)', in *Oxford Dictionary of National Biography* (Oxford: Oxford University Press, September 2004), online edn., updated October 2007, at http://www.oxforddnb.com/view/article/9320 (accessed 25 June 2010); and C. Davenhall, 'James Ferguson: a Commemoration', *Journal of Astronomical History and Heritage* 13, no. 3 (2010): pp. 179-86.

travelling showman and lecturer who worked at the margins of the same tradition. His performances were a mixture of genuine scientific phenomena and apparatus with conjuring tricks and other entertainment.

Fig. 2: 'The Kentish Hop Merchant and the Lecturer on Optics' (1809). This early nineteenth-century caricature illustrates the perils of the itinerant lecturer. During a talk in Kent about optics before a small audience a local merchant mishears the subject as 'hop-sticks'. The instruments on the lecturer's table include a telescope and a magic lantern. This hand-coloured engraving is by Isaac Cruikshank after a drawing by the caricaturist George Woodward. (Reference: 10198206; reproduced courtesy of the Science and Society Picture Library)

The first untethered, manned balloon flight took place on 21 November 1783 in a balloon made by the Montgolfier brothers and launched from near Paris. Balloon flights then became a craze that spread rapidly through Europe. They proved equally popular in the British Isles, with the first manned flight being made from Edinburgh on 27 August 1784. At the height of the balloon craze Katterfelto claimed not just to have made balloon flights, but to have conducted astronomical observations during them and to have made important discoveries from these observations. These claims are almost certainly fabrications: there is no evidence for such astronomical observations or discoveries and Katterfelto probably never made any balloon flights. This short note describes this curious, illusory prehistory to astronomical ballooning.

Fig. 3: A lunar halo and cross above the balloon *Zénith* in March 1875. The balloon was flown by the brothers Gaston and Albert Tissandier, who were pioneering meteorologists and aviators. Scientific, and particularly meteorological, ballooning did not take off until the mid-nineteenth century. The drawing is likely to be by Albert Tissandier. (Reference: LC-DIG-ppmsca-07435; courtesy the Library of Congress)

Actual astronomical ballooning began much later. Balloons began to be used as platforms for making scientific, particularly meteorological, measurements from the mid-nineteenth century (see Fig. 3) and occasional astronomical measurements were made from balloons from this time.[5]

[5] For meteorological ballooning see, for example, K. Anderson, *Predicting the Weather: Victorians and the Science of Meteorology* (Chicago: Chicago

Victor Hess established the extra-terrestrial origin of cosmic rays from balloon-borne measurements in 1912.[6] However, modern telescopic astronomical observations from balloons began in 1956 when Blackwell, Dewhirst and Dollfus photographed the sun.[7]

Gustavus Katterfelto

Gustavus Katterfelto's origins, the details of his life, even his date of birth and proper name, are unclear. His biography is further confused by his considerable capacity for self-aggrandisement and embellishing his own life-story. He was of German extraction, but other details of his early life remain obscure. The first facts that are known with reasonable certainty are that he and his wife arrived in Hull from Germany in 1776 and he almost immediately began advertising and presenting his show. It seems likely that he was already an experienced showman and had been travelling in Europe for some years. He claimed to have given a series of prestigious presentations before Royalty throughout the capitals of Europe, but these performances are almost certainly inventions.

Katterfelto's early years in Britain are also obscure, though he probably toured mostly in Northern England for a few years. In 1780 he moved to London, initially setting up for business in the Great Rooms at the Spring Gardens in the West End. Formerly a Huguenot chapel, by the 1780s the Spring Gardens had become a prestigious venue. Though Katterfelto demonstrated contemporary scientific apparatus, such as globes and orreries, his shows were unashamedly spectacle and entertainment and they quickly gained him a reputation. In 1782, London suffered a serious influenza epidemic on which he was able to capitalise. His equipment included a solar-illuminated microscope which he used to project images of the germs ostensibly responsible for the outbreak (but actually just random bacteria in a drop of water). He also sold patent remedies that he advertised as a sure cure for the contagion. Unsurprisingly, almost as

---

University Press, 2005), pp. 92, 96-97. An early example of astronomical observations from a balloon is J. Glaisher, 'Lines in the Solar Spectrum, as observed in the Balloon Ascent, 31st March last', *Monthly Notices of the Royal Astronomical Society* 23 (1863): pp. 191-92.

[6] D. Kolak, 'Hess, Victor Franz', in T. Hockey, ed., *Biographical Encyclopaedia of Astronomers* (New York: Springer, 2007), pp. 500-1.

[7] D. E. Blackwell, D. W. Dewhirst and A. Dollfus, 'Photography of Solar Granulation from a Manned Balloon', *The Observatory* 77, no. 896 (1957): pp. 20-23.

quickly as his reputation as a showman rose, he gained a counter-reputation as a quack and mountebank, peddling worthless remedies.

Katterfelto was also famous for his outrageously inflated self-publicity, or 'puffing' in the language of the time. He had handbills printed and placed advertisements in newspapers for his shows, all making extravagant claims for the wonders that would be presented. He would also pseudonymously submit to newspapers reviews of recent shows, ostensibly from satisfied customers and again extolling the wonders that were on display. There were also hints, which again he encouraged, of a darker side to his performance, with rumours that some of the effects were created using occult powers or with diabolic assistance. He kept a black cat (who subsequently produced a brood of black kittens) who was suspected to be his familiar.

Katterfelto's reputation declined as rapidly as it had risen. In 1784 he left London and from then until his death in 1799 he eked out a precarious living as a travelling showman, continuously touring the English provinces and Scotland. He is buried in the Church of St Gregory in Bedale, North Yorkshire.

Flights of Fancy

In late 1784 Katterfelto was touring East Anglia, having left London in July. At this time the balloon craze was at its height and Katterfelto was always keen to incorporate the latest sensation into his shows. While in London he had claimed to have launched unmanned balloons before Catherine the Great in St Petersburg as early as 1762. Had this outrageous and unsubstantiated claim been true it would have made him a genuine pioneer of ballooning. The following report appeared in the Norfolk Chronicle for 2 January 1785:

> The 23rd of December was the day that Dr Katterfelto was to ascend in his large Air Balloon in this city, with … various mathematical instruments etc. to take some astronomical observations. The day was very clear for it, but rather too cold to continue for a long time in the highest part of the atmosphere, so Dr Katterfelto, therefore did not expect it was in his power on that day to make observations sufficient, so he was obliged to put off his ascending; as he made his calculation by his thermometer, that if he had ascended three miles only from the earth that day, the cold must at that distance have been nine times greater, and he was to remain a few hours in the evening in the highest

> part of the atmosphere purposely for astronomical observation, he would therefore have felt the cold still more severe.[8]

It seems that Katterfelto was planning a balloon ascent from Norwich to make astronomical observations, but the attempt had to be postponed due to cold weather. A few days later a second report in the Norfolk Chronicle gave some further details:

> The London, Dublin, Edinburgh, Glasgow, Aberdeen, Oxford and Cambridge papers express that the philosophers and learned at the above cities have a very great desire of hearing from the Norwich, Bury and Ipswich papers that Dr Katterfelto has ascended in his air-balloon in the city of Norwich, as the learned gentlemen in the above cities do expect that Dr Katterfelto will make some very useful discoveries in Astronomy etc. etc. Dr Katterfelto being by all accounts one of the first astronomers as well as philosophers in the three kingdoms, and as the observations which he made four years ago at Greenwich have caused since that time a great advantage to this kingdom, particularly to the navy.[9]

Katterfelto was now claiming to have already made important ground-based astronomical observations some years earlier, as well as promising air-borne ones to follow. There is no evidence that any ascent was made from Norwich. These reports, however, presaged a pattern. Similar articles appeared in the local press wherever Katterfelto was touring throughout the following year. They were doubtless written by Katterfelto himself: it had long been his practice to put his own accounts in local papers in order to drum up interest in his shows. The reports often describe flights that he had made before Royalty in London and the discoveries that had followed. They only appear where Katterfelto was currently touring and are not mentioned in the London press or Court Circulars.

The series concludes with a report in the Lincoln, Rutland and Stamford Mercury for 30 December 1785 that an ascent had:

---

[8] *Norfolk Chronicle*, 2 January 1785, quoted in Paton-Williams, *Katterfelto*, p. 110.

[9] *Norfolk Chronicle*, 8 January 1785, quoted in Paton-Williams, *Katterfelto*, p. 118.

> … proved of very great benefit to our navigators and it is expected will in time benefit the whole world in general, and as no person in this kingdom, or abroad, have made any useful discoveries by the ascending in their various air balloons, besides that great and wonderful philosopher Doctor Katterfelto, the gentlemen belonging to the Admiralty, as well as the whole Royal Society, have made a report to the King, that a salary may be granted to the Doctor for his useful discoveries, and if it is only £300 a year, they think it is no more than he is worthy of...It is also reported that Dr Katterfelto is to be admitted at the next meeting a member of the Royal Society.[10]

By 1786 the balloon craze was over and Katterfelto dropped the subject. It is doubtful if he ever made a balloon flight.

Discussion

It is difficult to think of a worse platform for making astronomical observations than the rolling basket of an early balloon with, in the case of hot air balloons, the additional disadvantage of broiling, churning air from the brazier. Modern telescopic astronomical observations from balloons did not begin until the 1950s. Katterfelto's claims to have made astronomical discoveries from balloons are an illusory prehistory to astronomical ballooning.

Katterfelto certainly had some knowledge of astronomy. Globes, orreries and telescopes featured in his shows. In Liverpool in 1791 he used telescopes and other apparatus to project the solar eclipse of 3 April before an audience. His letters to various East Anglian newspapers in 1785 show an awareness of contemporary astronomical concerns. For example, the claimed improvements in navigation that would result from his work mirror the widespread interest in applying astronomical methods to navigation, particularly the determination of longitude at sea.

Katterfelto's invented claims to Royal patronage and election to the Royal Society seem an attempt to outdo the honours bestowed on his compatriot William Herschel following the then-recent discovery of Uranus in 1781. In particular, his claim of a recommendation for a pension of £300 per annum was, inevitably, more than the £200 per annum awarded to Herschel. Alternatively, as an itinerant lecturer and showman, perhaps he was trying to appropriate the sort of honours that had been

---

[10] *Lincoln, Rutland and Stamford Mercury*, 30 December 1785, quoted in Paton-Williams, *Katterfelto*, pp. 118-19.

bestowed on James Ferguson some twenty years earlier, when Ferguson was awarded a pension of £50 per annum and elected an FRS.

In any event, Katterfelto's claims echo the astronomical concerns and developments of his time. Particularly when touring in the provinces he was far from centres of fashion, culture and learning and was performing to a largely lay audience. Nonetheless he seems to have assumed that references to recent astronomical and scientific issues would find some resonance with his target public.

Acknowledgements

David Paton-Williams has recently published the first full-length biography of Katterfelto. His excellent study introduced me to Katterfelto and was the source for much of the material discussed here. The interpretation and any mistakes remain my own.

Celestial Art: An Interview with Geoff MacEwan

Nicholas Campion

September 2015 sees the next exhibition at the Joan Oliver 'Maneu' Galeria d'Art in Palma, Mallorca, of work by Geoff MacEwan, one of the most important British artists working in Spain. The exhibition will coincide with the annual 'Nit de l'Art', a major event in the Spanish artistic calendar. MacEwan has lived on the island of Mallorca since 1991, and shuttles back and forth between the UK and his Mediterranean home. He originally settled on the island when a fellowship from the Miró Foundation enabled him to work with Joan Barbera, Joan Miró's printmaker.

MacEwan's work is held in many public and private collections including The National Gallery of Modern Art in Edinburgh, the Fundació Joan Miro in Palma de Mallorca and the Contemporary Prints Department in The Victoria and Albert Museum in London.

I interviewed MacEwan in 2014 in his home in Soller, Mallorca, following the exhibition of his work at the Christ Church Picture Gallery in Oxford from 12 March to 5 May 2014. The exhibition was timed to coincide with the city's annual Times Literary Festival, and featured twenty-seven out of a total of forty-four prints inspired by Dante's *Divine Comedy*. As the exhibition notes told us, 'Dante's Divine Comedy has inspired artists for centuries; among them the British artist Geoff MacEwan, whose abstract visual interpretation of the text invites us to look and read again'.[1]

Dante's *Divine Comedy* occupies a midway point in western thought between the soul's ascent to the stars poetically described in Plato's *Republic*, and modern celestial journey literature from Jules Verne to

[1] 'Inferno – Purgatory – Paradise: Geoff MacEwan interprets Dante's Divine Comedy', Christchurch Picture Gallery, Oxford, 12 March–5 May 2014, at http://www.chch.ox.ac.uk/gallery/exhibitions/current [accessed 4 June 2014].

Nicholas Campion, 'Celestial Art: An Interview with Geoff MacEwan', ed. Nicholas Campion and Rolf Sinclair, *Culture and Cosmos*, Vol. 18 no. 1, Spring/Summer 2014, pp. 53-68.
www.CultureAndCosmos.org

Arthur C. Clarke. All share the proposition that wisdom is to be found in a journey to the stars. There have been several significant visual interpretations of the *Divine Comedy*, perhaps the most well known being Gustave Doré's. Over the past thirty years MacEwan has given Dante's its most radical, imaginative and impressionistic treatment.

NC: *Geoff, where did you train to be an artist?*
GM: First of all at Goldsmiths and then for two years at the Slade as a postgraduate. I was very lucky to have had my early training at Goldsmiths when Andrew Forge was principal. His approach was very open and not over controlling. We had Basil Beattie and Bert Irving as painting tutors and I also got valuable input from the sculpture department. At the Slade it was different.

NC: *In what way?*
GM: It was 1967 and respect for painting itself was being challenged. The Vietnam War had polluted all things American, the critic Clement Greenberg who had championed *avant garde* American Painting lost his authoritative hold, and a new critique started to emerge based more on philosophical and sociological investigation than pictorial innovation.

The concept of illustration was very much looked down on. We were painters and not illustrators! The painting was a thing in itself. This was a totally modernist conception, of course, and the people that were held up as exemplars were artists such as Mondrian, Pollock, Rothko and the abstract expressionists. Illustration was looked down on because anything that had a literary reference was regarded as second-rate. Yet because I was always interested in literature, there was always in me a desire to find a way to bring words and images together. When I went through a political phase, a lot of the texts I used were rousing passages from Marx, which I wouldn't use now. The student riots of '68 in Paris accelerated this process of critical fermentation. Boundaries between disciplines were being broken down. I'd been accepted by the college on the basis of my paintings, but in the two years I was there I didn't produce a single one.

***NC**: So what did you do?*
GM: I worked with two other students—Maggi Hambling and Harry Biggin—on a multi media project called *A Space of Five Times*.[2] We showed it at the Grabowski Gallery in South Kensington in 1969.

***NC:** And after that?*
GM: I had to get a job. By this time I was married with a small son. I took all sorts of work, but none of it paid enough. Then I got taken back by the Slade as a technical assistant and a year later went to work as a part-time lecturer at Falmouth School of Art. At the end of that year they didn't renew my contract. Once more I was out of work but meantime my marriage had folded. I decided to reinvent myself, so for the next five years I worked in the electronics industry, first as an illustrator and then as a writer. I had given up on Art and its Objects.

I was very confused and angry. I had thoroughly lost my way and everything I touched was tainted by bitterness. My work as a technical writer kept me occupied and in the end paid me very well, but it was a hollowed out existence. Finally I had to deal with my personal demons and only after that did I return to painting. I don't in the least regret this absence, it was a very important prelude to the next stage in my life. I came back to painting very committed and inspired.

***NC:** When did you decide to work from the Divine Comedy?*
GM: When I gave up my job as a writer I was living in a remote village twenty miles outside Edinburgh. My first exhibition was with a small but progressive gallery called the 369 Gallery. My work for that show consisted of large paintings on paper which also included text. Their imagery was drawn from the very wild landscape in which I was living and the texts were either homemade or else quotations from whatever I was reading at the time.

***NC:** OK. So now you were working with words as well as images.*
GM: Yes and I was very happy with this combination. The largest piece in that show was called *O clouds unfold*, from William Blake's words in 'Jerusalem', and was inspired by the Polish revolution and the movement of Russian troops up to the border in 1981—a moment of very high

---

[2] Hilary Whitney, 'theartsdesk Q&A: Artist Maggi Hambling' at http://www.theartsdesk.com/visual-arts/theartsdesk-qa-artist-maggi-hambling [accessed 4 Jun 2014].

tension. I managed to combine a map of Europe with a merged Landscape/Sunset, with Blake's words over-stenciled. The piece was 2 metres by 4 and flanked by 2 side panels containing quotations from Engels and Thoreau.

NC: *Do you have any photographs of this work?*
GM: Nowadays I could have taken plenty, but in those days it was problematic and I was strangely careless about these pieces. They were meant to be transitory, like wall posters in China. I was still in thrall to a politicized art, but without knowing how to achieve it. My decision to work from Dante was actually an attempt to resolve this dilemma.

NC: *There are different readings of Dante, obviously, but one is that he was working in a tradition in which the celestial journey brought the soul closer to God. How do you interpret that in your work and motivation?*
GM: Well, in this lies an irony because, by choosing to work from Dante, I was moving away from any direct social-political critique and was engaged instead with the conflicts within my own psyche. In fact I'd embarked on a spiritual journey.

NC: *You started with the Inferno. How did you go about it?*
GM: Page by page, just as if I was experiencing the poem as a journey in real time. These were paintings—oil on paper—and I worked quite quickly. I had a problem with choosing a translation. I found Henry Cary's translation unreadable so instead I chose a prose text by John Sinclair. There were 25 images in the series. I didn't impose meanings outside the text. Instead I allowed the poem to dictate the imagery and its treatment.

I dealt with it in a very sort of straightforward way. I simply read the poem and whenever I felt an image rise I painted. It was programmatic, and I didn't search for anything beyond what the text stimulated or gave rise to an image.

NC: *I hate labels but that's in the tradition of surrealism, in the sense of allowing the image to arise spontaneously.*
GM: I have always tried to avoid interfering with spontaneous images as they emerge during a painting. The biggest influence from my Art School days was Anton Ehrensweig with whom I only exchanged a handful of words. His 1953 book *The Hidden Order of Art* made a great impression on me and other artists of my generation. It taught me to accept confusion

as a necessary stage in the process of arriving at an image and its completion. In other words, it taught me not to be fearful.

NC: *So, if you're reading the words and an image arises inside you, does that image come from just you as an individual? I'm asking because there's a school of thought that claims that the image arises from the collective unconscious. Have you ever thought about that? Do you think the image is yours and you are able as an individual to create a meaningful image which then speaks to other people? Or, if an image arises, does it arise from something universal.*

GM: There are obviously collective archetypes, symbols that recur time and again throughout art history. I arrive at my images through the process of painting. They come into being as a result of attempting to resolve the technical problems of composition because, for me, painting is structured dialogue. I am continuously sensitive to what is happening on the canvas; watchful for those footprints which will carry the painting forward. Hunting and tracking are good metaphors for my way of working. As for whether I think of my images as specifically mine: they have come into being as a result of my actions but once they are visible and have been given a context, they belong to the collective of images.

Over the years certain images have made a regular appearance in my work, and it isn't because they have very precise meanings for me but because at some point in the process they add symbolic weight to a painting, a weight by association.

My painting at the time was quite expressionistic, and at the same time linked with natural forms. I was very straightforward in the sense that I tried to make each image cope if possible: cope symbolically but also cope in a particular way, carrying the weight of the ambiance as if we would go down into the inferno, itself.

For example, the pomegranate is a very beautiful fruit, and has associations for me with autumn in London and a first love affair. I know the story of Persephone and four seeds—pomegranate seeds—that she was tricked into eating by Hades, and how that explains the seasons. But, above all, I love the fruit for its beauty, the acid yellow pith, the delicate colours of the tightly packed seeds and the russet peel. It stands for life and plenty.

So when I produced the paper work *Night Flight*, based on Bush's bombing of Afghanistan, I included a pomegranate as a contradictory presence and later, in a series of etchings called *Cascaras* I drew the twisted husk of the empty fruit as a symbol of our mortality.

NC: *The pomegranate already had an archetypical identity but, as you pointed out, it had intimate associations for you. So when you inserted it into your drawing it was loaded with a whole range of meanings, and not just private ones.*

GM: The whole business is complicated in the sense that if you've been an artist for any length of time you will have looked at a lot of art and absorbed a great deal of it. When I'm working I'll often be reminded of another person's work in a chance coincidence of brushwork. So I have memories of what I've seen before and may even use them. For instance, the image of Beatrice at the end of the *Purgatorio* is based very loosely on Blake's *Beatrice Addressing Dante from the Car*, which was part of the Tate's collection that I've known since I was a student.

The process produces the shadow of something and I guide it into being. Some artists say that they feel they are a channel through which something flows from outside themselves into their work. The pieces that succeed are those in which this transubstantiation isn't blocked.

But what is interesting and was something that I did not determine but came about quite naturally, was that the first illustration in the *Purgatorio* (Canto 1) (Fig. 1) was very structured. It's a very structured linear piece and in black and blue.

And then the final image (Fig. 2, *The Earthly Paradise*) is totally loose and dynamic and all the constraint has gone, and because it's an earthly paradise, it's in a vegetable form, and it's also like a dance. So you can see that, as the *Purgatorio* unfolds, the structures become progressively looser. Gradually there is a moving away from the initial and oppressive structure.

NC: *Your first work based on Dante was a series of paintings that you exhibited at the 369 Gallery in Edinburgh. What year was that?*

GM: 1982. Andrew Brown, the Gallery's Director, was very supportive. He produced a fine Letter-press catalogue with an introduction by Jonathan Usher from the University's Italian Department. I sold the whole set of paintings privately several years later.

Fig. 1: *Purgatorio* Canto 1 (*The Reed Bed*).

Fig. 2: *Purgatorio* plate 15 (*The Earthly Paradise*).

NC: *When did you start the printed editions of the Divine Comedy?*
GM: In 1990 I was commissioned by Edinburgh University Library to create a limited edition of 21 etchings based on the Inferno. Just after that project ended I left Scotland and then there was a gap of twenty years before Joan Oliver Maneu in Mallorca supported my production of the *Purgatorio*. The plates based on the *Paradiso* followed three years later.

NC: *Twenty years is a long time. You obviously worked on many other pieces.*
GM: Yes. I was very productive. A series of paintings, works on paper and latterly several folios of prints. Four one man shows.

NC: *Did you have access to a print studio?*
GM: I didn't have a press in Mallorca until eight years ago, so I had to go to Madrid and work with a very accomplished printer there. I did four projects with Dan Benveniste; all of them turned out really well. His technical ability is phenomenal and he has the sensitivity and intellectual rigour of an artist.

I produced *El Proceso de Ramon Lull* in Madrid. It's the finest piece that I've ever produced and the perfect example of what I was talking about earlier. It simply flowed into being—not without a lot of effort of course. And there was a serious moment that confirmed Anton Ehrensweig's psycho-analytical approach to creativity.

NC: *When was this?*
GM: 1994. I was introduced to Dan Benveniste be the then curator of Prints at La Reina Sofia. I wanted to work on a very large plate and Dan's studio had a huge press. My idea was to etch a plate and print the edition; then add more to the plate and print again. We would take the plate through six stages and end up with six complete sets. There was the danger of messing up half way through, but that was exciting

Along the way we ran into a Resistance. By the fourth stage I was so worn out - I'd been in Madrid for six weeks - that I wanted to end the project on the fifth stage. The fourth print was very dark indeed and I wanted to conclude the series by physically scraping the whole plate clean. Dan disagreed. Five in a series was awkward. Two-one-two is not ideal. Also, with very little work on the plate the overall blackness could be subtly scraped away to reveal its hidden structure. I held out all afternoon. And come the evening we were back on track.

But why had I stood in the way of the complete unfolding of the image at its most crucial stage? Because the progress of this work demanded that the darkness should be analyzed and clarified—only then could the final print have its redemptive meaning. In this incident I recognized a deep-rooted carelessness that had either spoilt or distorted many of my actions, artistic or otherwise.

Fig. 3: Ramon Lull 1.

Fig. 4: Ramon Lull 4.

NC: So the artist himself often stands in his own light?
GM: It's often the case. That particular instance was very dramatic. Even now, years later, I feel a little like someone who almost deserted and had to be persuaded to stay.

***NC**: Coming back to Dante, the first image that really struck me was that of* The Penitents *from your version of the* Purgatorio, *with the faces stitched together.*
GM: The eyes sewn up. Horrible, but at least they have the consolation of an eventual release and their suffering is an act of reconstruction. And as well as reparation for their Envy. In the previous plate the painter Oderisi da Gubbio is being purged of Pride and Dante uses him as a justifier of his own poetic work.[3]

Oderisi, the medieval illuminator, has felt the sting of injured pride but he now realizes that there is a natural progression in Art and gives a little homily on the subject mentioning Giotto and Dante as those who are in forefront of the new style. Oderisi is a sort of alter ego for me. I've felt a little like him from time to time.

Dante placed Oderisi in the Circle of Pride. It's interesting because it's a little bit of art history. Oderisi was what would be called a Paris illuminator. In other words, he would be considered old-fashioned by someone, say, like Giotto, who Dante mentions in the same passage. Oderisi was still essentially medieval and feels he hasn't been given due credit for this work. But, as a result of being in purgatory, he explains to Dante that art is a moveable feast. It moves on and, as a result, some people get left behind. First one person is holding the torch, then another. Then he too will fall back and someone else will take his place. Of course, Dante uses this passage as an opportunity to push his new style of writing, using Oderisi as an example of the person who is left behind. And so Oderisi has always been quite an important figure for me, left behind as he is, in the *Purgatorio*.

In the *Purgatorio* I tried to bring lightness to the judgmental structure. Unlike the Inferno, where there is no repentance but only the claustrophobia of repeated anguish, the penitents are rising through proscribed punishments towards their release.

In the first Plate of the series—*The Reed Bed*—the image of an angel's wing is hemmed in by vertical columns representing the Hell that Dante has left behind, and the mountain of Purgatory that he will have to climb. It's a tightly disciplined composition. Only the blue gives it any lightness. The final plate—*The Earthly Paradise*—shares the same blue; this was a deliberate linkage, otherwise the image is the exact opposite; a free-flowing and floral celebration of release.

---

[3] The thirteenth century painter Oderisi da Gubbio. Dante, *Purgatorio*, XI, pp. 79-123.

NC: *From what?*
GM: Guilt.

Fig. 5: *Pride* (*Purgatorio* plate 9).

Fig. 6: *Envy* (*Purgatorio* plate 8).

NC: *How deeply did you become involved in the religious philosophy of the poem?*
GM: As I said earlier, I came to the poem hoping to resolve the problem of art's relation to politics; after all, the *Comedia* is full of political theory and invective, so I could have taken a contemporary fix and found present day equivalents; but instead I became involved in the dramas of the various individuals that haunt the circles of hell and the terraces of purgatory.

You'll perhaps think it strange when I tell you that I believed everything that I read, that I was convinced by Dante's grand design and believed in its redemptive power. This is something that art can do. It can break through the carapace of ennui and intellectual cynicism. What do you feel when you listen to the final chorus of Bach's St. Matthew Passion? I think awe gets close to what I feel.

NC: *Is this what art is for you? A means whereby you can approach the unsayable—the awesome?*
GM: A few years ago I was in Madrid working on a project with Dan and went as usual to the Prado to revisit my favourite paintings. They have an altar piece there by Van Weyden which depicts Christ being taken down from the cross. It's a powerful and acknowledged masterpiece which draws you into a drama of grief and disappointment. Like all great works it holds you in thrall. Later on the same day I went to an exhibition of videos by Bill Viola and watched a piece called *Emergence 2002*.[4] This twelve-minute video, involving three actors, was a dramatization of a deposition and an enactment of grief that was inspired by a fifteenth-century fresco by Masolino da Panicali. Both the painting and the video, through their extreme precision of construction, provide the stage for emotion to unfold.

In fact, the more I gazed at Van Weyden's painting, the more I was drawn into an emotional composition where every detail supported the sad ceremony. Nothing was surplus to the pure intention of the piece and there could be no distraction from my involvement in their grief.

In the case of Bill Viola's video, which was a projected experience, every second of the unfolding drama was intensified by the slow orchestration of the actors. The emergence of the Christ figure from the font, the spilled water and the movements of his mother and the other Mary are so filled with the dignity of love that once again nothing is wasted or stands in the way of catharsis. One of the problems of contemporary art is

[4] https://www.youtube.com/watch?v=FagLc3rOV88 [accessed 4 June 2014].

how, in an atmosphere of irony and knowingness, to create the opportunity for deep emotional involvement.

NC: *In terms of the literary background to Dante, there are literal ideas of the soul's ascent as a journey through the spheres of the planets to salvation, or to the divine, or self-realization. In Dante's time there was a pronounced idea that that journey is not a literal one, but takes place inside.*
GM: For me the journey itself was crucial, as were Dante's portraits of the damned and the penitents. When you talk about salvation I would interpret that as the resolution of all inner contradictions. *Statius Redeemed*—Canto 25—deals with the poet's release from Purgatory. The conflicting portions of his personality have been finally resolved, but the equilibrium reached is still an earthly one. The eight *Paradise Plates* deal with what happens after. It's essentially a post-mortem process.

NC: *I feel that you have a special interest in this section, over and above what has gone before. Am I right?*
GM: When I first showed the Dante prints at Ca'n Prunera in Soller I was very pleased to see how well all three sections interacted. I realized that I had been right to etch the *Paradise* plates in the way that I had, and that they'd ended my journey on exactly the right note.

The *Inferno* was executed in dry-point, totally consistent with the dramas of that section. The *Purgatorio* required a measure of restraint so the etches were never very deep or extensive and were softened by colour; but the *Paradise Plates* were heavily worked on from the beginning, not just with acid and aquatint but also with extensive scraping away and burnishing. The text for this section was unwritten. There was no effort to translate; everything evolved from moment to moment, from day to day, week to week. Don't ask for meanings, I told myself, just do it.

NC: *How long did it take you?*
GM: A long time. Almost six months. But I was so absorbed in their genesis that I hardly noticed, even when there were setbacks. For me it was like working on a large painting. Very intense. I would stand for hours looking at a plate, running my finger over its surface, feeling the lines and textures as if I was a blind man in search of a landscape.

NC: *Did you find one?*
GM: In the end I found an archetypal sequence and something I hadn't expected, but which was consistent with everything that had gone before. I always felt this last section had to reflect the didactic tone of the *Paradiso*. That's why I chose Canto 2, where Beatrice explains the distribution of Divine energy and why the Moon is stained in the way that it is as the subject matter for the first plate. After that the rest of the series followed on very easily and naturally.

NC: *These prints have a very cosmic feel to them.*
GM: I know and it really feels as though immense forces are at play here that are barely describable. It reminds me of the passage from Joseph Glanville that Poe quotes as a heading for his story 'A Descent into the Maelstrom', and which sums up my attitude to all things metaphysical. My idea was that the end result of the *Purgatorio* was that paradise was reached, and to that extent the internal contradictions had finally been resolved. And so now, the *Paradiso* begins after the resolution of the contradictions. My idea is that in the *Paradiso*, we are talking about what death might actually mean, what might come after death.

NC: *And what did you get to with that thought?*
GM: Well, I got to the *Paradiso* plates, which are meditations on that process. The interesting thing is that, when seen as a whole, they actually are very consistent—a consistent unfolding which leads to a final image (*Paradiso* Plate 8, Fig, 7), which is a form of rebirth. But I wanted to give them a cosmic feel, and I think they are very powerful from that point of view.

If you take the penultimate image (*Paradiso* Plate 7, Fig. 8), we have here, for me, the lighthouse, which for me is to search the truth and meaning. But it's also the thing that guides you home—guides the sailor home.

But the point about the seas is that the form within which the lighthouse sits is transferred to the final plate (*Paradiso* Plate 8), and it becomes the body of a woman in cross-section. What lies at the centre is something that is embryonic in all its beginnings of a new life, which ties in with the first illustration (*Paradiso* Plate 1), which looks very much like the fertilisation of the egg.

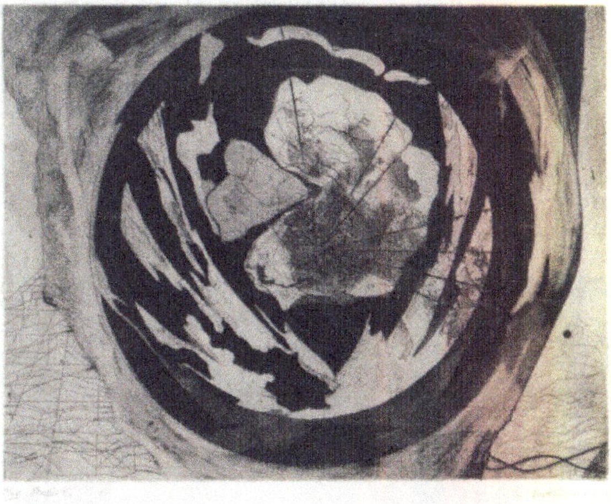

Fig. 7: *Paradiso* Plate 8.

Fig. 8: *Paradiso* Plate 7.

NC: With my assumptions I saw Plate 1 as a planet or a comet.

GM: Well, it can be whatever you want it to be! The first plate was an attempt to deal with Beatrice's very complicated description of how God's grace flows downwards through the heavenly spheres, and it's about the moon. Dante comes up with this optical experiment. But when you look at it, you've got these little things moving towards it, so there is a penis and some sperm. I didn't conceive it like that but one can see it like that. And so, where we had one, now the one has become two. This was not consciously thought of at all. It was merely a consequence of my attempt to be cosmic, and to portray energy flowing through the universe.

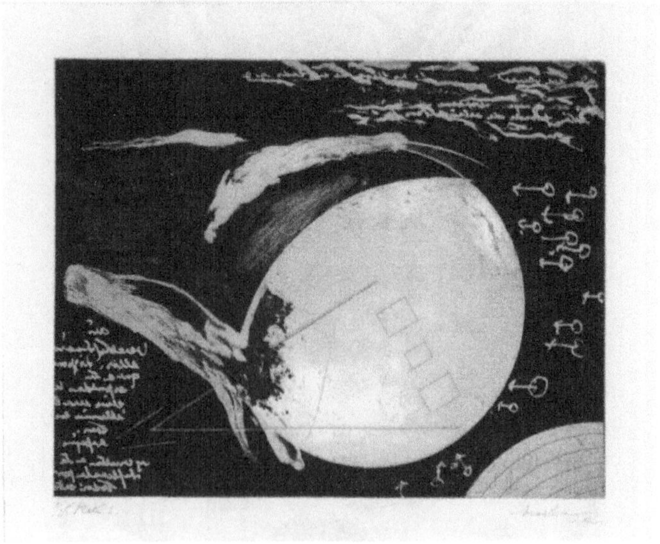

Fig. 9: *Paradiso* Plate 1.

NC: *We haven't had time to discuss your other pieces, some of which deal with existential themes such as identity and the nature of consciousness. What are you working on now?*

GM: I want to work from nature again, that's the best way to enter a new phase. But right now I am preparing new paintings for my exhibition in Palma in September.

NC: *Thank you.*

For Geoff MacEwan's work online see:
http://geoffmacewan.blogspot.com.es/
and
http://www.sinclairspress.com/geoffmacewan

NOTES ON CONTRIBUTORS

César Esteban has a PhD in Astrophysics. Lecturer at the University of La Laguna and researcher at the Instituto de Astrofísica de Canarias. Although his main research topics are the chemical composition of ionized nebulae and chemical evolution of the Universe, he devotes part of his time to archaeoastronomy. He has carried out fieldwork in the Iberian Peninsula, Canary Islands, North Africa, islands of Polynesia and Micronesia and the Valley of Mexico. In recent years, he has developed research projects on Tartessian and Iberian sanctuaries and necropolises in south and eastern Spain in collaboration with archaeologists.

Ronald Hutton is Professor of History at the University of Bristol (August 1996 to date). Prior to this he was a Fellow of Magdalen College, Oxford, before serving as Lecturer and then Reader in History at Bristol (1981–96). He is also the historian and prehistorian on the Board of Trustees which runs English Heritage, and chair of the Blue Plaques Panel. Since the 1980s he has been involved in the writing and presentation of documentaries for various television channels. He is a Fellow of the Royal Historical Society, the Society of Antiquaries, the Learned Society of Wales and the British Academy. His many published works include *The Stations of the Sun: A History of the Ritual Year in Britain* (1996); *The Triumph of the Moon: A History of Modern Pagan Witchcraft* (1999); *Blood and Mistletoe: The History of the Druids in Britain* (2009); and *Pagan Britain* (2013).

Nick Kollerstrom was a member of staff of University College London for eleven years, from 1997 to 2008. He was elected as a Member of the New York Academy of Sciences in 1999, is a Fellow of the Royal Astronomical Society, was a founder-member of the Society for History of Astronomy and has had a couple of dozen articles published in academic journals, on history of astronomy, here: http://dioi.org/kn/index.htm

Clive Davenhall has a long-standing interest in the history of astronomy. Since 2004 he has been the Editor of the Society for the History of Astronomy's *Bulletin* (previously *Newsletter*) and has contributed entries for the *Biographical Encyclopaedia of Astronomers*. In real life he is a Project Manager and Software Developer in the Wide Field Astronomy Unit, Institute for Astronomy, University of Edinburgh, based at the Royal Observatory Edinburgh.

BACK ISSUES OF CULTURE AND COSMOS

http://www.cultureandcosmos.org/backIssues.html

Contents, Vol. 1 no 1 (spring/summer 1997)
Robin Heath: *An Astronomical Basis for Solar Hero Myths;* **Norris Hetherington**: *Ancient Greek Cosmology and Culture: a Historiographical Review;* **Alan Weber**: *The Development of Celestial Journey Literature, 1400 - 1650;* **Ken Negus**: *Kepler's Tertius Interveniens;* **John Durant** and **Martin Bauer**: *British Public Perceptions of Astrology: an Approach from the Sociology of Knowledge.*

Contents Vol. 1 no 2 (autumn/winter 1997)
Otto Neugebauer: *On the History of Wretched Subjects;* **Nick Kollerstrom**: *The Star Zodiac of Antiquity;* **Robert Zoller**: *The Hermetica as Ancient Science;* **Edgar Laird**: *Christine de Pizan and Controversy Concerning Star Study in the Court of Charles V;* **Jürgen G.H. Hoppman**: *The Lichtenberger Prophecy and Melanchthon's Horoscope for Luther;* **Elizabeth Heine**: *W.B.Yeats: Poet and Astrologer.*

Contents Vol. 2 no 1 (spring/summer 1998)
J. McKim Malville and **R. N. Swaminathan:** *People, Planets and the Sun: Surya Puja in Tamil Nadu, South India;* **Carlos Trenary:** *Yaxchilan Lintel 25 as a Cometary Record;* **Graziella Federici Vescovini:** *Biagio Pelacani's Astrological History for the Year 1405*; **Frank McGillion:** *The Influence of Wilhelm Fliess' Cosmobiology on Sigmund Freud;* **Nicholas Campion:** *Sigmund Freud's Investigation of Astrology.*

Contents Vol. 2 no 2 (autumn/winter 1998)
Giuseppe Bezza: *Astrological Considerations on the Length of Life in Hellenistic, Persian and Arabic Astrology;* **Angela Voss:** *The Music of the Spheres: Marsilio Ficino and Renaissance harmonia*; **Robert Zoller:** *Marc Edmund Jones and New Age Astrology in America.*

Contents Vol. 3 no 1 (spring/summer 1999)
Michael R. Molnar: *Firmicus Maternus and the Star of Bethlehem*; **Roger Beck:** *The Astronomical Design of Karakush, a Royal Burial Site in Ancient Commagene: an Hypothesis*; **Chantal Allison:** *The Ifriqiya Uprising Horoscope from* On Reception *by Masha'alla, Court Astrologer in the Early 'Abassid Caliphate.*

Contents Vol. 3 no 2 (autumn/winter 1999)
Robin Waterfield: *The Evidence of Astrology in Classical Greece;* **Remo Catani:** *The Polemics on Astrology 1489-1524*; **Claudia Rousseau**: *An Astrological Prognostication to Duke Cosimo de Medici of Florence.*

Contents Vol. 4 no 1 (spring/summer 2000)
Patrick Curry: *Historical Approaches to Astrology*; **Edgar Laird:** *Heaven and the Sphaera Mundi in the Middle Ages*; **George D. Chryssides:** *Is God a Space Alien? The Cosmology of the Raëlian Church.*

Contents Vol. 4 no 2 (autumn/winter 2000)
David J. Ross: *The Bird, The Cross, and the Emperor: Investigations into the Antiquity* of *The Cross in Cygnus*; **Angela Voss:** *The Astrology of Marsilio Ficino: Divination or Science?;* **Patrick Curry:** *Astrology on Trial, and its Historians: Reflections on the Historiography of 'Superstition'.*

Contents Vol. 5 no 1 (spring/summer 2001)
Demetra George: *Manuel I Komnenos and Michael Glykas: A Twelfth-Century Defence and Refutation of Astrology,* Part I; **Richard L. Poss:** *Stars and Spirituality in the Cosmology of Dante's* Commedia.

Contents Vol. 5 no 2 (autumn/winter 2001)
Arkadiusz Sołtysiak: *The Bull of Heaven in Mesopotamian Sources*; **Demetra George:** *Manuel I Komnenos and Michael Glykas: A Twelfth-Century Defence and Refutation of Astrology,* Part 2; **Garry Phillipson** and **Peter Case:** *The Hidden Lineage of Modern Management Science: Astrology, Alchemy and the Myers-Briggs Type Indicator.*

Contents Volume 6 Number 1 (spring/summer 2002)
Ari Belenkyi: *A Unique Feature of the Jewish Calendar - Deĥiyot*; **Demetra George:** *Manuel I Komnenos and Michael Glykas: A Twelfth-Century Defence and Refutation of Astrology*, Part 3; **Germana Ernst**: *The Sky in a Room: Campanella's Apologeticus in defence of the pamphlet* De siderali fato vitando; **Tommaso Campanella:** *Apologia for the opuscule on* De siderali fato vitando.

Contents Volume 6 Number 2 (autumn/winter 2002)
Jesse Krai: *Rheticus' Poem* 'Concerning the Beer of Breslau and the Twelve Signs of the Zodiac'; **Anna Marie Roos:** *Israel Hiebner's Astrological Amulets and the English Sigil War*; **Nicholas Campion:** *Surrealist Cosmology: André Breton and Astrology.*

Contents Volume 7 Number 1 (spring/summer 2003) GALILEO'S ASTROLOGY
Nick Kollerstrom: *Foreword: Galileo as Believer*; **Nicholas Campion**: *Introduction: Galileo's Life and Work*; **Antonio Favaro**: *Galileo, Astrologer*; **Germana Ernst**: *Astrology and Prophecy in Campanella and Galileo*; **Nick Kollerstrom**; *Galileo as an Astrologer: Antonino Poppi: On Trial for Astral Fatalism: Galileo Faces the Inquisition;* **Guiseppe Righini**:*Galileo's Horoscope for Cosimo II de Medici*; **Mario Biagioli**: *An Astrologico-Dynastic Encounter; Galileo's Correspondence*; *Galileo's Letter to Dini, May 1611*; *On the Character of Sagredo: Galileo's judgements upon his nativity*; *Galileo's Horoscopes for his Daughters*; *Rome, 1630*; **Bernadette Brady**: *Four Galilean Horoscopes: An Analysis of Galileo's Astrological Techniques*; *A Sonnet by Galileo.*

Contents Volume 7 Number 2 (autumn/winter 2003)
Günther Oestmann: *Tycho Brahe's Geniture*; **Bernard Eccles**: *Astrological physiognomy from Ptolemy to the present day*; **James Brockbank**: *Planetary signification from the second century until the present day*; **Julia Cleave**: *Ficino's Approach to Astrology as Reflected in Book VII of his Letters.*

Contents Volume 8 No 1/2 (spring/summer autumn/winter 2004)
Valerie Shrimplin *Organising INSAP*; **Rolf Sinclair** *Foreword: INSAP IV in Oxford: A Summary*; **Nicholas Campion** *Introduction: The Inspiration of Astronomical Phenomena*:

Hubert A. Allen, Jr. *Hawkins' Way: Remembering Astronomer Gerald S. Hawkins*; **Hubert A. Allen, Jr. and Terry Edward Ballone** *Star Imagery in Petroglyph National Monument*; **Mark Butterworth** *Astronomy and the Magic Lantern*; **Ann Laurence Caudano** *Sun, Moon, and Stars on Kievan Rus Jewellery (10th – 13th Centuries)*; **Nicholas Campion** *The Sun is God;* **Anne Chapman-Rietschi** *Cosmic Gardens*; **Deborah Garwood** *Paris Solstice*; **N. J. Girardot** *Celestial Worlds In the Work of Self-Taught Visionary Artists With Special Reference to Howard Finster's Vision of 1982*; **John G. Hatch** *Desire, Heavenly Bodies, and a Surrealist's Fascination with the Celestial Theatre*; **Holly Henry** *Bertrand Russell in Blue Spectacles: His Fascination with Astronomy*; Ronald Hicks *Astronomy and the Sacred Landscape in Irish Myth*; **Chris Impey** *Why Are We So Lonely?*; **Bernd Klähn** *The Aberration of Starlight and/in Postmodernist Fiction*; **Nick Kollerstrom** *How Galileo dedicated the moons of Jupiter to Cosimo II de Medici*; **Arnold Lebeuf** *Dating the five Suns of Aztec cosmology*; **Andrea D. Lobel** *Trailing the Paper Moon: Astronomical Interpretations of Exodus 12:1-2*; **Stephen C. McCluskey** *Wordsworth's 'Rydal Chapel' and the Astronomical Orientation of Churches*; **David Madacsi** *Sky: Atmospheres and Aesthetic Distance in Planetary and Lunar Environments*; **Daniel R. Matlaga** *A Journey of Celestial Lights: The Sky as Allegory in Melville's Moby Dick*; **Paul Murdin** *Representing the Moon*; **R. P. Olowin** *Robinson Jeffers: Poetic Responses to a Cosmological Revolution*; **David W. Pankenier** *A Brief History of Beiji (Northern Culmen)*; **Richard Poss** *Poetic Responses to the Size of the Universe: Astronomical Imagery and Cosmological Constraints*; **Barbara Rappenglück** *The material of the solid sky and its traces in cultures*; **Brad Ricca** *The Night of Falling Stars: Reading the 1833 Leonid Meteor Storm*; **Patricia Ricci** *Lux ex Tenebris: Etienne-Louis Boullée's Cenotaph for Sir Isaac Newton*; **Sarah Richards** *Die Planetentheorie: its uses and meanings for the Saxon mining communities and the culture of the Dresden Court 1553-1719*; **William Saslaw and Paul Murdi**n *The Double Apollos of Istrus*; **Petra G. Schmidl** *Dusk and Dawn in Medieval Islam; On the Importance of Twilight Phenomena with Some Examples of Their Representations in Texts and on Instruments*; **Valerie Shrimplin** *Borromini and the New Astronomy: the elliptical dome*; **Joshua Stein** *Cicero's Use of Astronomy as Proof of the Existence of the Gods*; **Antje Steinhoefel** *Art and Astronomy in the Service of Religion:Observations on the Work of John Russell (1745-1806)*; **Burkard Steinrücken** *An interpretation of the 'Sky Disc of Nebra' as an icon for a bronze age planetarium mechanism with parallels to the moving world-soul in Plato's* Timaeus; **Gary Wells** *Daumier and The Popular Image of Astronomy.*

Contents Vol. 9 no 1 (Spring/Summer 2005)
Gennadij Kurtik and Alexander Militarev *Once more on the origin of Semitic and Greek star names:an astronomic-etymological approach updated*; **Prudence Jones** *A Goddess Arrives: Nineteenth Century Sources of the New Age Triple Moon Goddess*; **Louise Curth** *Astrological Medicine and the Popular Press in Early Modern England.*

Contents Vol. 9 no 2 (Autumn/Winter 2005)
Marinus Anthony van der Sluijs *A Possible Babylonian Precursor to the Theory of ecpyrōsis*; **Liz Greene** *Did Orphic Beliefs Influence the Development of Hellenistic Astrology?*; **Ariel Cohen** *Astronomical Luni-Solar Cycles and the Chronology of the Masoretic Bible*; **Tayra Lanuza-Navarro** *An Astrological Disc from the Sixteenth Century*; **J.C. Holbrook** *Celestial Navigators and Navigation Stories.*

Contents Vol. 10 no 1 and 2 (Spring/Summer, Autumn/Winter 2006)

Lucia Dolce *Introduction: The worship of celestial bodies in Japan: politics, rituals and icons*; **Lucia Dolce** *The State of the Field: A basic bibliography on astrological cultic practices in Japan*; **Hayashi Makoto** *The Tokugawa Shoguns and Yin-yang knowledge (onmyōdō)*; **John Breen** *Inside Tokugawa religion: stars, planets and the calendar-as-*method; **Mark Teeuwen** *The imperial shrines of Ise:An ancient star cult?*; **Lilla Russell-Smith** *Stars and Planets in Chinese and Central Asian Buddhist Art from the Ninth to the Fifteenth Centuries*; **Matsumoto Ikuyo** *Two Mediaeval Manuscripts on the Worship of the Stars from the Fujii Eikan Collection*; **Tsuda Tetsuei** *The Images of Stars and Their Significance in Japanese Esoteric Buddhist Art*; **Meri Arichi** *Seven Stars of Heaven and Seven Shrines on Earth: The Big Dipper and the Hie Shrine in the Medieval* Period; **Gaynor Sekimori** *Star Rituals and Nikko Shugendô*; **Meri Arichi** *The front cover image: Myōken Bosatsu.*

Contents Vol. 11 no 1 and 2 (Spring/Summer, Autumn/Winter 2007)
Micah Ross *A Survey of Demotic Astrological* Texts; **Francis Schmidt** *Horoscope, Predestination and Merit in Ancient Judaism*; **Stephan Heilen** *Ancient Scholars on the Horoscope of Rome*; **Joanna Komorowska** *Philosophy among Astrologers* ; **Wolfgang Hübner** *The Tropical Points of the Zodiacal Year and the* Paranatellonta *in Manilius'* Astronomica; Aurelio Pérez Jiménez *Hephaestio and the Consecration of Statues*; **Robert Hand** *Signs as Houses (Places) in Ancient Astrology*; **Dorian Gieseler Greenbaum** *Calculating the Lots of Fortune and Daemon in Hellenistic Astrology*; **Susanne Denningmann** *The Ambiguous Terms* ἑῴα *and* ἑσπερία, ἀνατολή, *and* ἑῴα *and* ἑσπερία δύσις **Joseph Crane** *Ptolemy's Digression: Astrology's Aspects andMusical Intervals*; **Giuseppe Bezza** *The Development of an Astrological Term – from Greek* hairesis *to Arabic* ⍰ayyiz; **Deborah Houlding** *The Transmission of Ptolemy's Terms: An Historical Overview, Comparison and Interpretation.*

Contents Vol. 12 no 1 (Spring/Summer 2008)
Liz Greene *Is Astrology a Divinatory System?*; **James Maffie** *Watching the Heavens with a 'Rooted Heart': The Mystical Basis of Aztec Astronomy*; **J.C. Holbrook** *Astronomy and World Heritage.*

Contents Vol. 12 no 2 (Autumn/Winter 2008)
Mark Williams *Astrological Poetry in late medieval Wales: the case of Dafydd Nanmor's 'To God and the planet Saturn'*; **Scott Hendrix** *Choosing to be Human: Albert the Great on Self Awareness and Celestial Influence*; **Graham Douglas** *Luis Vilhena and the World of Astrology.*

Contents Vol. 13 no 1 (Spring/Summer 2009)
Josefina Rodríguez-Arribas *Astronomical and Astrological Terms in Ibn Ezra's Biblical Commentaries: A New Approach*; **Andrew Vladimirou** *Michael Psellos and Byzantine Astrology in the Eleventh Century*; **Marinus Anthony van der Sluijs** *The Dragon of the Eclipses—A Note*; **Patrick Curry** *Response to Liz Greene's 'Is Astrology a Divinatory System?'*

Contents Vol. 13 no 2 (Autumn/Winter 2009)
Liz Greene *Mystical Experiences Among Astrologers*; **Peter Pesic** *How the Sun Stood Still: Old English Interpretations of Joshua and the Leap Year*; **Doina Ionescu** *Virginia Woolf and Astronomy*; **Carlos Ziller Camenietzki and Luis Miguel Carolino** *Astrologers at*

Contents Vol. 14 no 1 and 2 (Spring/Summer, Autumn/Winter 2010)

Contents Vol. 15 no 1 (Spring/Summer 2011)

Contents Vol. 15 no 2 (Autumn/Winter 2011)

Contents Volume 16 No 1/2 (Spring/Summer Autumn/Winter 2012) *of Astronomy*; **Euan MacKie**, *A new look at the astronomy and geometry of Stonehenge*; **Leonid Marsadolov**, *Archaeoastronomical Aspects of the Archaeological Monuments of Siberia*; **Christian Etheridge**, *A systematic re-evaluation of the sources of Old Norse astronomy*; **Aidan Foster**, *Hierophanies in the Vinland Sagas: Images of a New World*; **Inga Elmqvist Söderlund**, *Inspiration from antique heroic deeds: Hercules as an astronomer*; **Patricia Aakhus**, *Astral Magic and Adelard of Bath's Liber Prestigiorum; or Why Werewolves Change at the Full Moon*; **David Pankenier**, Astrology for an Empire: The 'Treatise on the Celestial Offices' (ca. 100 BCE); **Steven Renshaw**, *The Inspiration of Subaru as a Symbol of Values and Traditions in Japan*;b **Daniel Armstrong**, *Citing The Saucers: Astronomy, UFOs and a persistence of vision*; **Alberto Cappi**, *The concept of gravity before Newton*; **Paul Murdin**, *Artilleryman to head of state—how astronomy inspired Francois Arago*; **Paolo Molaro and Alberto Cappi**, *Edgar Allan Poe's cosmology in* Eureka; **Voula Saridakis**, *For 'the present and future happiness of my dear Pupils'": The Astronomical and Educational Legacy of Margaret Bryan*; **Michael Rowan–Robinson**, *The invisible universe*; THE ARTS: **Arnold Wolfendale**, *The Inter-Relation of the Visual Arts and Science in General and Astronomy in Particular*; **Lynda Harris**, *Changing Images of the Milky Way during the Greco-Roman and Medieval Periods*; **Lucia Ayala**, *The Universe in images: Iconography of the Plurality of Worlds*; **Tayra M. Carmen Lanuza-Navarro**, *Astrological culture before its public: the representation of astrology in Golden Age Spanish Theatre*; **Emily Urban**, *Depicting the Heavens: The Use of Astrology in the Frescoes of Rome*; **Michael Mendillo**, *The Artistic Portrayal of the Medicean Moons in Early Astronomical Charts, Books and Paintings*; **Rolf Sinclair**, *Howard Russell Butler: Painter Extraordinary of Solar Eclipses*; **Beatriz Garcia, Estela Reynoso, Silvina Pérez Alvarez and Rubén Gabellone**, *Inspiration of Astronomy in the movies: a history of a close encounter*; **Gary Wells**, *The Moon in the Landscape: Interpreting a Theme of 19th Century Art*; **Clive Davenhall**, *The Space Art of Scriven Bolton*; **Matthew Whitehouse**, *Astronomical Organ Music*; **Aaron Plasek**, *Between Scientists, Writers and Artists: Theorising and Critiquing Knowledge-Production at the Interstices between Disciplines*; ARTISTS: **Merja Markkula**, *The Way I See the Stars: fibre art inspired by astrobiology*; **Govinda Sah**, *Beyond the Notion*; **Gisela Weimann**, *Above all the stars*; **Courtney Wrenn**, *Nebulae (emission / absorption)*; **Toby MacLennan**, *Presentation of Playing the Stars*; **Felicity Spear**, *Extending vision: sky-situated knowledge and the artist's eye*; **Vanessa Stanley**, *Surveillance-Surveillance-Surveillance*; **Jim Cogswell**, *Molecular Delirium.*

Contents Vol. 17 no 1 (Spring/Summer 2013)
Clifford J. Cunningham and Günther Oestmann *Classical Deities in Astronomy: The Employment of Verse to Commemorate the Discovery of the Planets Uranus, Ceres, Pallas, Juno, and Vesta*; **Dorian Knight** *A Reinvestigation Into Astronomical Motifs in Eddic Poetry*; **Karen Smyth** *'I specially note their Astronomie, philosophie, and other parts of profound or cunning art': The Use of Cosmos Registers by Chaucer and Others*; **Kirk Little** *Spellbound: The Astrological Imagination of Washington Irving*; **Guiliano Masola and Nicola Reggiani** *Σελήνη Τοξότῃ: Business and Astrology in the Papyri*; **Reinhard Mussik** *Research Note: Weltall, Erde, Mensch and Marxist Cosmology in East Germany*

Contents Vol. 17 no 2 (Autumn/Winter 2013)
Daniel Brown: *The Experience of Watching: Place Defined by the Trinity of Land-, Sea-, and Skyscape*; **Pamela Armstrong:** *Skyscapes of the Mesolithic/Neolithic Transition in Western England*; **Olwyn Pritchard,** *North as a Sacred Direction?*

Traces of a Prehistoric North-South Route Across Pembrokeshire; **Tore Lomsdalen:** *The Islandscape of the Megalithic Temple Structures of Prehistoric Malta*; **Fernando Pimenta, Nuno Ribeiro, Anabela Joaquinito, António Félix Rodrigues, Antonieta Costa and Fabio Silva:** *Land, Sea and Skyscape:Two Case Studies of Man-made Structures in the Azores Islands*

Lightning Source UK Ltd.
Milton Keynes UK
UKOW06f0212260316

270920UK00011B/46/P

9 781907 767715